In the Presence of Nature

For Theresa

David B. Wilson

In the Presence of Nature

David Scofield Wilson

UNIVERSITY OF MASSACHUSETTS PRESS   AMHERST, 1978

Library of Congress Catalog Card Number 77–90733
ISBN 0–87023–020–4
Printed in the United States of America
Designed by Mary Mendell
Library of Congress Cataloging in Publication Data
appear on the last printed page of the book

Illustrations courtesy of The Bancroft Library.
Frontispiece: "The Globe Fish. *Orbis Lævis Variegatus*"
(volume 2, plate 28) from Mark Catesby's *Natural History
of Carolina* (London, 1771) in the collection of The
Minneapolis Athenaeum, 300 Nicollet Mall, Minneapolis,
Minnesota 55401. Photograph by Tom Didiaso.

Part of chapter 5 originally appeared in "The Iconography
of Mark Catesby," *Eighteenth-Century Studies* 4:2 (Winter
1971), 169–83. Copyright © 1971 by The Regents of the
University of California. Reprinted from *Eighteenth-
Century Studies* by permission of The Regents.

For Bonnie

Contents

Illustrations

Introduction

As a study of a particular subject—nature, nature reporting, and
the transmission of beliefs from one world to another—this
book speaks for itself, needing little introduction. *In the Presence of Nature* is a study of cultural transmission, of what happens to a body of values when it is transported from a familiar
surround to an unfamiliar one. Such works form a staple of
scholarship in American culture studies, serving—among other
things—to articulate how America relates to the Old World,
and to itself. A quick inventory of studies in and around this
genre would include: Henry Nash Smith's classic *Virgin Land,*
Leo Marx, *The Machine in the Garden,* R. W. B. Lewis, *The
American Adam,* Kenneth Lockridge, *A New England Town,*
Philip Greven, *Four Generations,* Richard Slotkin, *Regenera-
tion through Violence,* Daniel Boorstin, *The Americans: The
Colonial Experience,* John William Ward, *Andrew Jackson:
Symbol for an Age,* Charles Sanford, *The Quest for Paradise,*
Perry Miller, *The New England Mind: From Colony to Prov-
ince,* Kai Erikson, *Wayward Puritans,* Howard Mumford Jones,
O Brave New World, Darrett Rutman, *American Puritanism.*
David Wilson's *In the Presence of Nature* deserves a place
among these notable works.

At first glance, it might not seem so. For what distinguishes
the studies of a Smith, a Marx, a Lewis, a Miller is not simply
the quality of their scholarship, but their choice of such momentous subjects. Smith sweeps over a century and a half of America's cultural history, from the Revolution to the first World
War, updating Turner on the relation between inherited idea
and native thing in the New World. Lewis is concerned with
the basic urge of Americans to identify themselves, exploring
that urge in masters of American literature—Thoreau, Melville,
Hawthorne et al. Slotkin penetrates the American disposition to
violence, both symbolic and actual, again spanning centuries in
his work. Ward adheres to a more limited time frame, but fo-

cuses on one of the half-dozen giants of the American presidency, and what that character reveals about the culture of a people. Sanford, not limiting himself to the mundane particulars of time and place, tackles the whole symbolic meaning of the modern world, and America's role in that world. Lockridge and Greven, contending that historic meaning is revealed *through* the mundane particulars of time and place, use those particulars to try and recapture "a world we have lost," defining the modern by contrasting for us life in a premodern surround. Miller wrestles with the Puritans wrestling with their own titanic furies—God, Salvation, the Devil, and the relations of Man to same.

If nature on occasion is titanic in David Wilson's book, certainly his human subjects are less so. Here we find no Melvilles, no Mathers, no Jacksons, but Jonathan Carver, John Bartram, Mark Catesby—three rather undistinguished figures of the eighteenth century. What gives the book its power as scholarship, then, if not the inherent power of its subjects? That question is the focus of this introduction; it is the kind of question an author cannot easily answer of his own work, but an outsider may. It is an issue not so much of subject, where—as I have noted—the book speaks for itself. It is an issue rather of method, or approach to subject.

My answers to this question come on several different levels. The first is intimately connected to Wilson's subject matter, but is not inherent in it. It lies rather in how the author addresses his subject. Fundamentally, the book is distinguished by Wilson's ability to make nature—or rather the confrontation of received belief with nature—*problematic* through the figures of Carver, Bartram, and Catesby. That did not happen by accident. Nor did it happen simply by going to all the documents and reading them carefully. It happened by going to the documents *armed with* a well-articulated theory of how problematic is the experience of people; that theory is guided by what sociologists Peter Berger and Thomas Luckmann call the inherent "world-openness" of the human situation.

Wilson writes that men, common men as well as exceptional, *make* the universe; they inhabit and alter or maintain it by their consent or resistance. Here are not block beliefs, then, not finished bodies of symbol and myth and image; here rather are incomplete human worlds in the making and other incomplete worlds in the disassembling. In this sense, David Wilson's strategy resembles that of Kenneth Lockridge and Philip Greven; he seeks to rediscover "a world we have lost"—not in his case a world of farm and village, but a world of unhuman things and beings with which today's cosmopolitans have lost familiarity. As Wilson says, "It is difficult for modern readers to recapture the problematic air of American nature for the early reporters. It was not a matter simply of an unfamiliar plant here, a slightly different snake there. The whole design of American nature seemed unsettling, as if anything might happen when one left civilization."

If Wilson's book is not peopled by momentous subjects, then, it is powered by a momentous problem—the problem of an awe-ful, uncertain, and unexplored world requiring from humans close observation first of all, then some form of itemizing, then finally credible explanation. It is this confrontation between experience and explanation—of small men addressing strange subjects in a strange world, and grasping for a hedge against radical uncertainty—which gives *In the Presence of Nature* its elemental drama.

That is why David Wilson calls his figures "nature reporters." They are reporters not only in the obvious sense that, as Wilson says, they "searched for news of nature." They are reporters also in the more fundamental sense that, as agents of cultural communication in a changing human order, "they transformed the cosmos by making the nature of the new world . . . familiar to all who could read or be read to." That is, they investigated, reported on, and attempted to explain problematic experience in their day. In the case of Carver, Bartram, and Catesby in the eighteenth century, the problematic was preeminently in nature, just as the problematic in our day seems preeminently in and around the political. It is no accident, then, that as nature reporters occupied an important communicative role in eighteenth-century America, so political reporters play a central communicative role in contemporary America. One should resist facile comparisons, especially with a work as refined as Wilson's, but perhaps the twentieth-century cultural analogues to Jonathan Carver, John Bartram, and Mark Catesby are not Joseph Wood Krutch and Ansel Adams and

Sand County Almanac, but Gary Wills (*Nixon Agonistes*), Frances Fitzgerald (*A Fire in the Lake*), Theodore White (*The Making of the President*), David Halberstam (*The Best and the Brightest*), Bernstein and Woodward (*All the President's Men*). Culturally, these reporters help people *locate* themselves in their times by investigating what is most problematic in their collective experience, and by trying to explain what is strange and troubling and sometimes terrifying in their world.

How *In the Presence of Nature* was created gives us a second explanation of its intellectual power. A careful reader will soon detect what David Wilson tells us later in his epilogue; this book was a long time in the making. Informally, as Wilson says, the book began when he was a child, stimulated by a native curiosity about nature and things. Formally, research started when he was a graduate student in American Studies at Minnesota, and the project was provisionally completed as a dissertation in 1968. Now, a full ten years later, the project is published in book form.

Had it been published soon after it was finished as a dissertation, this study would doubtless have plugged gaps in our understanding of eighteenth-century America, would thus have been counted as a solid contribution to knowledge, and would thereby have satisfied academic promotion and tenure committees of Professor Wilson's "promise as a creative and productive scholar." But it would not have resulted in the book—or anywhere near the book—we see here.

For this book shows the refinement, the maturity, the intellectual compression of a manuscript thought and rethought, worked and reworked over the years, rather than one researched under pressure and published prematurely to satisfy the external demands of scholarly "productivity." One does not under pressure write sentences like: "Carver was a type as well as an individual. We are all, and that is the drama of biography"; or, "Culture defines the cosmos, but not for all time and not neutrally"; or, "I take this juxtaposition of nausea and spouting to be deliberate and wry, but do not press the interpretation"; and, "John Bartram is a problem. . . . Unlike . . . Carver . . . he is hard to catch." Sentences like this come not at first thought, nor second, nor third; they come from one grown familiar with a subject over the years, familiar enough to relax with it, play

with it, subject it to the progressive reconsiderations of one's
own mind and experience.

We find in Wilson's book, then, a deliberately understated
wisdom about life and things. *In the Presence of Nature* is thus
a genuinely *creative* work of scholarship; it not only offers us
new things to see and know, it creates fresh ways to see and
know them, and provides a rich context to do it in. In this sense,
the book far transcends the superficial academic's sense of "creative"—meaning simply to make a contribution to knowledge.
In this sense also it indicts the trend in academe forcing scholars
to publish before they are ready, and bears witness to the wisdom of allowing a mind to set its own intellectual pace rather
than holding it to arbitrary schedules.

Another quality accounting for the book's power is David
Wilson's writing style—or rather his thinking style expressed in
words. Next to R. W. B. Lewis's *American Adam, In the Presence of Nature* is the most intensely crafted book I know of
those associated with the American Studies movement. Wilson
wastes hardly a word in this volume; it is as carefully formed
as it is carefully researched. Wherever the reader looks, there is
evidence of the author's *mind;* Wilson never lets the pressure
of undigested data overwhelm him. Most of his sentences are
lean and pure, but when long they move with a rhythm which
fits the subject matter.

In the Presence of Nature is not only a wise book, then, it is
a disciplined book. That too comes with time. Discipline prematurely imposed may degenerate into excessive caution and
constriction. But discipline allowed to mature in its own time
can result in the fresh, creative forms by which a mind can
grow to control its subject matter.

Finally, we can help explain the intellectual power of David Wilson's book by locating it in traditions of recent American scholarship. *In the Presence of Nature* is pre-eminently the
creation of an individual vision; yet it also connects to an ongoing community of works in American culture studies, and
addresses matters of concern to that community.

During the "golden years" of American Studies scholarship,
from 1950 to around 1965, culture studies tended to mean the
study of *words*. And that further meant words penned mostly
by men of letters and by professional men—ministers, occasion-

ally lawyers or doctors, philosophers and historians and such. There were exceptions of course, as there are to any general practice. Henry Nash Smith looked at dime novels and political tracts as well as at works of high literature; John William Ward studied responses to Andrew Jackson in popular songs and cartoons. Yet in virtually all the notable American Studies works of this decade and a half, scholars approached their materials from a humanistic bent. They were most concerned with how Americans expressed themselves verbally; they were less concerned with how Americans behaved beyond the verbal, and they were minimally concerned with social institutions that function below the level of verbal expression.

After 1965, a reaction set in. Empowered by what was called "the new social history," historical scholars sought to split off the social and behavioral from the study of ideas and expression. Works like Philip Greven's *Four Generations*, for example, wholly ignored what men and women in colonial Massachusetts Bay thought—even hesitating to call them "Puritans"—and concentrated instead on size of family, land-holding patterns, birth and death rates, and so on.

American culture studies, as traditionally conceived, was beleaguered by this new development, and for a time the sources of integrative scholarship in and around American Studies threatened to dry up. Only in recent years have we found indications of a new cultural synthesis—a fresh convergence of the intellectual and the social which acknowledges criticisms from new social history but is not overwhelmed by them. Robert Sklar's *Movie-Made America* is a model of this convergence, as is Thomas Haskell's *The Emergence of Professional Social Science*. Both works are steeped in humanistic perspectives—Sklar from literature and film studies, Haskell from philosophy and intellectual history. But both also convey respect for the integrity and the power of social institutions. What distinguishes Sklar and Haskell is how they handle transactions *between* ideas and social structure, between what is normally called intellectual history and social history. In this respect, they serve as exemplars of a new kind of culture studies, a fresh integrating synthesis for the late 1970s.

David Wilson's *In the Presence of Nature* also indicates this convergence. Like Sklar and Haskell, Wilson approaches culture studies from the humanities. As I have noted, a major virtue of the book is his keen sensitivity to the uses of words, and the beauty of his own usage. Yet traditional humanists will encounter a strange language here. We read not of genres and symbols and images so much as of "culture work" and "plausibility structures" and "recipe knowledge" and "cognitive maps" and "ethnoscience." If Wilson's subject is traditional humanist-literary, many of his tools are drawn from contemporary social science, especially from the cognitive sociology of Peter Berger and Thomas Luckmann. Wilson's book stands firmly in the humanities, then, but its tools of analysis build a bridge to the social sciences. It is, to be sure, the cognitive and expressive side of social science, that side closest to the humanities. Unlike Sklar and Haskell, Wilson is less concerned with institutional forms than with expressive forms; this is not a criticism so much as a notation of what further scholarship in the area—more thoroughly grounded in social structural concerns—might pick up on.

In the Presence of Nature is not only a fine book by itself, then. It joins a handful of works published in the last few years which offer hope for a rejuvenated kind of American culture studies—a culture studies building on the solid foundation of past humanistic scholarship but using perspectives from social science to transcend limitations in that scholarship.

Gene Wise
American Studies
University of Maryland

In the Presence of Nature

I

The Nature Reporter in the
Presence of Nature and Culture

Consent and Resistance

Nature is present to naturalists the way God is to saints or the past is to humanists—not simply as a matter of fact but as an insistent and live reality. To come face to face with a flying spider or a rattlesnake in the road unhinges habit and intensifies awareness, just as stumbling upon an ancient rune does, or encountering a burning bush. These uncommon phenomena throw the settled world into disarray. They seem to signify forgotten truths or webs of meaning until now invisible. Everyday meaning and value become problematic. Close attention must be paid and care taken if one is to make sense of it all, and if the world is to make sense still. There are some who are dead to the impact of such events, but for those who are not, every detail seems significant and its *gestalt* feels compelling. It is like the difference between those who encounter a poem as an "affecting presence" and those who see it only as a batch of words, perhaps documenting a mental state, but otherwise empty of special worth.[1] Like the difference that separates lovers of music from those insensitive to its special qualities; those unaffected may acknowledge the existence of the poem or the rattlesnake but still wonder what all the shouting is about.

Why some people see and feel a presence others miss is almost impossible to explain satisfactorily, but *how* they consent individually to a presence or deny its force (or even existence) is a matter of record. And how diverse persons manage to share particular values and meanings across time and space and across the gap between personalities goes to the heart of being human and to the core, then, of culture studies. How people differ or agree on what is and what matters makes a difference in the quality of life, day by day—esthetically, ethically, politically, in every way. And it makes a difference in the long run as well, since patterns set up in one age come down in the next. Being human depends on what we make of things and what we take as real and care about, especially as it all gets woven into "patterns-in-experience," or living culture.[2]

In the twentieth century naturalists have often been discounted as harmless hobbyists and would-be scientists. That was not always the case. In the eighteenth century they were taken more seriously and deemed worthy of institutional support or deserving of public correction. Today it takes an act of imagination and some acquaintance with the work they did to appreciate the contribution naturalists made to American culture. And it takes some familiarity with the world of eighteenth-century Europe and America to be able to place their work in a context that makes the human drama come alive. A useful entry into the ferment of conflicting values and meanings, of contrary assumptions and contending self-evident truths that enveloped early naturalists, is provided by Dr. Samuel Johnson's admonitory prescription. A poet does not "number the streaks of the tulip," he declared, "or describe the different shades in the verdure of the forest":

'The business of the poet,' said Imlac, 'is to examine, not the individual, but the species; to remark general properties and large appearances. . . . He is to exhibit in his portraits of nature such prominent and striking features as recall the original to every mind, and must neglect the minuter discriminations, which one may have remarked and another have neglected, for those characteristics which are alike obvious to vigilance and carelessness.' [Chapter X, *Rasselas* (1759)]

It makes sense, he implied, to speak of violets, honey bees and pines, but if a writer specifies stamens or catalogues conifers few will understand, and fewer yet appreciate. He believed that such minute discriminations betray the observer's industry more than they do nature's beauty.

The world of Dr. Johnson was not the world of Linnaeus. *Nature* meant one thing to humanists, another to naturalists. The humanists' world had evolved in history and incorporated ancient philosophy and Christian values and perspectives: man at the center, God at the top and nature subordinate to both. This cosmology justified a stance *vis-à-vis* the world that was at odds with the posture implicit in the cosmos the naturalists were groping to articulate. For them nature was central, and they wrote *natural* history and *natural* philosophy which challenged traditional beliefs. Sir Isaac Newton had already revolutionized physics and astronomy in the seventeenth century, and comets had become natural signs of cosmic order and harmony instead of portents of plague or signs of God's wrath. In the eighteenth century Benjamin Franklin tamed lightning and charted the Gulf Stream. Others sought to naturalize earthquakes and volcanoes. Physicians introduced inoculation to tame smallpox. And travellers reported natural herbal cures for syphilis and "discovered" Indian remedies unknown to Europe. Both humanists and naturalists saluted the motto *dulce et utile,* but as troops in the field they pursued divergent campaigns and foraged in different groves.

Tradition and Revolution

Distinguishing between science and the humanities, art historian Erwin Panofsky writes, "while science endeavors to transform the chaotic variety of human records into what may be called a cosmos of nature, the humanities endeavor to transform the chaotic variety of human records into what may be called a cosmos of culture." "Cosmos of nature" is an odd phrase at first glance. *Cosmos* or *nature* alone would seem enough, but Panofsky means to stress the inherent finitude and ultimate artificiality of any socially constructed cosmos, even the cosmos defined by scientists. To endorse Panofsky's conception one need not abandon faith that somewhere behind projected realities lies an ultimate actuality. Or that in their pursuit of truth, humanists and scientists equally seek to make their fictions congruent to "the facts." And yet humanists *qua* humanists and scientists *qua* scientists live in different worlds— and do so on purpose. In this regard, science and the humanities are competing culture projects.[3]

Panofsky, of course, writes from a twentieth-century perspective and perceives cultural patterns with the advantage of hindsight. Earlier students of nature and art did not always know they competed—indeed many strove to create coherent world views in which the natural and humanistic complemented each other. Many habitually and easily divided their energies between both endeavors or pursued projects that were at once both humanistic and natural. Newton saw no conflict between his physics and his theology. Jonathan Edwards embraced the "new science" of his day and enriched his theology with it. Franklin's urbanity and wit found expression in his descriptions of whirlwinds and the art of swimming as well as in his bagatelles and almanac pieces. And yet, individually successful accommodations notwithstanding, it has become clear to succeeding generations that the proper study of mankind and the proper study of nature became antithetical endeavors in a cultural dialectic.

The older tradition is that of the humanists. Long before the advent of the "new science" of the seventeenth century, humanists had accumulated and produced a considerable body of literature and art. Theory, methodology and techniques had been rationalized so that genres were distinguished, modes defined, styles described and criticism established. Humanism found institutional expression in the universities, societies, archives and libraries of Europe. The technology of printing advanced the distribution of texts and critiques, and by the eighteenth century periodicals made humanism a remunerative profession. Coffeehouses and soirees provided public and private arenas in which professional and amateur humanists could meet to confer, dispute and plan projects. In short, humanism had become established.

While humanism had harbored various students of nature from its beginnings and had bred many of the earliest fathers of science—men like Bacon and Newton—many humanists began to express an uneasy conservatism toward the end of the seventeenth century when they saw natural philosophers presume to employ the press, exploit the libraries and expropriate analytical logic. The humanists caught the scent of a revolution under way, and they took to the presses to expose "the moderns" and support what they believed to be right values and true meanings. Meanwhile, the glacially slow drift of western culture from sacrality to secularity intensified and accelerated under the pressure of new discoveries and emerging economic and social systems.

The "cosmos of nature" took centuries to develop its full challenge. Its roots fed on Christian and even pre-Christian notions that existence shows a bipolar design, a dialectic of spirit and matter, the ideal and real, the sacred and profane. It owes its thrust and distinctive plausibility, however, to the complex of economic, military, political and literary activity that attended its emergence. Early voyagers retrieving rare natural intelligence from faraway places lent support to the conviction that nature was an exciting and rewarding alternative environment to that furnished by civilized society. Merchant capitalists promoted the perception of nature as commodity, often fragmentary and exotic, when they searched "all corners of the new-found World / For pleasant fruits and princely delicacies" to import and sell; geographers delimited *terra incognita;* poets imagined new Edens and Arcadias; adventurers purveyed compelling stories of noble savages who held herbal and psychic knowledge long lost to western man—all contributed in one way or another to building the conviction that nature was a palpable and external reality which if exploited enriched the spirit and pocket of man.[4]

The revolution that eventuated in the transformation of the renaissance cosmos of culture into the modern cosmos of nature influenced and was influenced by other revolutions in western culture—the Protestant, the capitalist, the nationalistic and the democratic. But unlike these others it did not often employ or engender physical violence against human beings. It directed men's energies outward toward external and generally nonhuman nature rather than inward toward existing populations and power establishments. It was a revolution of human perspective—like the Copernican revolution—more than a seizure of power. Of course transfers of power occurred and enmities spilled over into verbal violence. Establishments were threatened, categories disarranged, decorums violated and gains reapportioned in much the same way, but not with the same physical agony to persons, as attended other revolutions of modern times.

The flux and flow of this long cultural revolution appears to have involved a shift of legitimacy, so that new men with different interests seized the right to define reality, gather instruments, control the terms and rhetoric of serious dialogue and exposition, institute societies and establish networks for the flow of material and ideas. The humanists and theologians had filled this role in the past; naturalists challenged and eroded their control in the seventeenth and eighteenth centuries; and scientists and technologists increasingly consolidated their triumph over the humanists and theologians in the nineteenth and twentieth centuries. Yet few revolutions are as total as regicide. Generally the old order fails to die, lives on as a part of the new and explains, rearranges, challenges and adjusts to the new conditions. So it was even in the 1960s with the talk of "two cultures"; so it was even more vitally in the seventeenth and eighteenth centuries when the new naturalism first seriously challenged the old order.

Probably actors in cultural revolutions never fully comprehend the import of their actions. Certainly a sixteenth-century voyager collecting riches from the new world could not foresee the effect on later ages of his exploitive behavior. Nor could a seventeenth-century optical technician, or an eighteenth-century chemist. In fact if one thing is clear from the history of science it is that even those men who have fathered new disciplines with their inventions, discoveries or syntheses of paradigms fail to anticipate all cultural reverberations of their deeds. Nor could they if they wished, for as a movement develops and changes it tends to establish new arrangements and initiate activities which in their turn alter conditions and foster still newer developments.

Instituting the Nature Project

With the establishment of the Royal Society of London (1660), England won control over much of the public debate in natural philosophy. The Royal Society legitimated the pursuit of natural truth and centralized the collection, organization and distribution of useful knowledge. Furthermore it generated two complementary role types: the correspondent and the initiated Fellow of the Royal Society. Other nations established similar societies (e.g., The Swedish Royal Academy of Sciences, The French Académie Royale des Sciences, The American Philosophical Society) and they too distinguished between Fellows and correspondents. The Royal Society of London was more like a mercantile company than a humanist university. It commissioned projects, patronized far-flung "seed factors," collected and stored imported natural products and exported digests of knowledge and symbolic rewards (medals, honorary posts, species named in one's honor). Many of the participants in the enterprise were merchants, colonial administrators, planters and missionaries—persons who in one way or another were tied to an imperial network of communication and flow of authority.

Various sorts of men and women are organized by any such system—mercantile, imperial, holy or natural—and nature study demanded many in the field to travel, observe, collect and report, while others, generally stationed in London, organized and systematized the raw material collected. Little formal discipline or special training was required of those in the field; consequently they seldom achieved rewards in their own time or recognition in a later time commensurate with their expenditure of energy and intellect. Yet they too, along with data and theories, were organized by the system.

Those who contributed to the growth of useful, natural knowledge have been called by many names. Not until the nineteenth century were they called *scientists,* a term that connotes a degree of specialization and organization not aptly applied to those of earlier centuries. Medieval and Renaissance men who studied nature's ways and products were called *surgeon, apothecary, astrologer, alchemist, navigator* or *herbalist* if their relationship to nature were somehow professionally legitimated and narrowly defined; or *theologian, cleric, witch, humanist* or *prince* if their cultural niche were larger and their nature study only a portion of the role defined for them by society. By the seventeenth and eighteenth centuries, certain nature studies had become sufficiently distinct from the old in purpose and method to demand new terms for those who studied nature in the new mode. They were called, and indeed called themselves, *natural philosophers, natural historians, virtuosi* or *curiosi*. While each term has its utility and nuances of meaning and connotation, none adequately distinguishes between those who systematized and rationalized the new knowledge about nature and those who observed and collected natural specimens. Many of the latter cannot satisfactorily be called *natural philosophers* or *natural historians,* though they occasionally speculated and synthesized as the natural philosophers and historians did. Their role is best described by the phrase *nature reporter;* their work, as *nature reportage.*[5]

The nature reporter searched for news of nature, for incidents that challenged or confirmed expectations, and for the myriad facts that natural philosophy took to be its concern. Nature reporters, like modern journalists—with whom they share a remarkable number of qualities—pursued their mission both abroad and at home. They worked free-lance or under direction and patronage, and dispatched raw information, dramatic and entertaining anecdotes, or digests of natural experience to the centers of science and the public at large. Occasionally they even perpetrated hoaxes on the gullible or overeager, as reporters are fond of doing. More often than not they were self-educated in the methods and techniques of their craft and only marginally or selectively acquainted with the philosophy and theory that justified and rationalized their activity. No less than natural philosophers, these reporters sought truth and felt truth to be beautiful, but they tended to find satisfaction in the concrete and in simple arrangements rather than in abstract concepts and formal systems. And they tended to find truth in simple observation, personal experience and useful natural products. They consistently reported the what, when, where and who of occurrences and often leaped quickly to guesses of why, for they were not disinterested observers but in part promoters of nature. They were the journeymen of the emerging natural history establishment and exhibited the qualities such a role often encourages and produces.

The technology of printing, which had developed with the humanities and had served them well, came increasingly in the seventeenth and eighteenth centuries to serve naturalists too. First, voyages and travels, then guidebooks, catalogues, epistles, periodical series and book-length natural histories full of minutiae and explanation flowed from the presses and attracted public patronage. Nature reporters also employed or adapted to new uses many of the conventional humanistic genres and modes. The most common were the journal, the essay, the epistle and the travel narrative. Additionally, however, the diary, anecdotal calendar, type sketch, woodcut, copper engraving and especially the sermon—with its pattern of text, explication and application or corollary—became common forms of reportage. Even the long narrative and descriptive poem was employed to transmit natural knowledge.

Cosmos Maintenance

Given the naturalists' appropriation of traditional forms, it is hardly surprising that established humanists often found the literary results raw, grotesque, unbalanced or otherwise disturbingly undisciplined and regarded the authors as illegitimate interlopers and charlatans at worst, as unrefined and uninstructed literary novices at best. They reacted as the established often do when they begin to lose control of the means of production and distribution—much as modern professional scientists often do when confronted with popularizations of their special knowledge and techniques. They strove with earnest logic, extravagant ridicule and urbane wit to "correct" the authors' ideas, tastes, values, styles and ethos. Or alternatively, they reflexively shot tidbits of exotic nature into their own otherwise conventional pastorals, lyrics, plays, satires, essays and tales.

Especially in the eighteenth century when humanists avidly sought cultural harmony and balance did men like Jonathan Swift, Addison and Steele, Alexander Pope, Samuel Johnson and Joshua Reynolds strive to maintain the cosmos of culture against modern enthusiasm and error. A common mode was ridicule. Swift characterized the "projectors" of the Royal Society as fools in Book III of *Gulliver's Travels*. In *The Battle of the Books* he likened them to filthy spiders, while comparing humanists to bees: the latter wandered far in sunlight to extract honey and wax— "sweetness and light" —from flowers for humanity's benefit; the former hid in libraries and spun fortifications out of excrement. According to Boswell, Dr. Johnson once replied when asked if he were a botanist, "No, Sir, I am not a botanist; and, should I wish to become a botanist, I must first turn myself into a reptile." Elsewhere Johnson satirized the virtuosi by portraying one who, as a child, had broken the playthings his mother gave him so that he "might discover the method of their structure, and the cause of their motions"; as an adult, this fool's passion for nature's rarities caused him to accept rent payments from his tenants in worms and insects so that his cabinet might boast "three species of earth-worms not known to the naturalists" and "the longest blade of grass upon record." [6]

An alternate mode of cosmos maintenance was adopted by Johnson and Sir Joshua Reynolds when they prescriptively addressed their audiences. Reynolds told his auditors at the Royal Academy of Arts (1776) that the proper balance between the general and particular in painting required them to adopt "the general idea of Nature" and avoid "exact representation of individual objects with all their imperfections," for to call such "particular living objects" *nature,* he said, was to misapply the term. To number the streaks of the tulip was for them both an epistemological and esthetic mistake.[7]

These humanists spoke the truth as they saw it—the truth of the eighteenth-century cosmos of culture. In the magazines and almanacs of the age, essayists employed types such as Tom Folio *(Tatler* 158), Sir Andrew Freeport *(Spectator* 2), Silence Dogood *(New-England Courant)* and Mrs. Busy *(Rambler* 138) to educate the public in the principles of economics, criticism, morality and taste. Poets adapted classical forms to modern subjects so that nightingales appeared as "the feathered kind," and abstract moral virtues, as trains of personifications, the loveliest maid of which is Poetry. Even in portraiture intended to convey the likeness of an individual, too close a literalness to contemporary dress was judged undesirable by Reynolds for it revealed, he said, the work of the tailor and not the essential man beneath. For the eighteenth-century humanist, art that had survived the revolutions of ages provided the standard of excellence, and to follow the ancients' forms and conventions, modes and decorum, was to approach again the original truth of nature. "Nature and Homer were ... the same," Pope declared. That is to say, nature meant laws and principles, not local and particular details.[8]

Making New Connections

The nature reporters marched to a "different drummer." They sought less to maintain an established order than to build a new complex of truth and beauty. Theirs was the younger enterprise and one for which they often felt it necessary to adopt new modes of thought, categorization and verification. The ancients may indeed have laid the groundwork for natural knowledge, but there was much they had missed, had failed to understand or had credulously taken for fact when it was fiction. These new men and women did more than number the streaks of the tulip. They pursued natural knowledge by questioning accepted formulations and starting deliberately at the bottom again to collect all the raw data they could. They would first get the facts, then construct theories to account for and order them.

Predictably, however, theory and practice did not always coincide. People are impatient and time is short, so while collecting surprising, delightful or confounding bits of particular natural knowledge, the natural philosophers frequently speculated at once, generated new theories, dismantled old ones and constructed fresh orders. So too did the nature reporters, for though the developing institutional structure of natural philosophy implied a division of intellectual labor between those in the field who were to collect and those at home who were to synthesize, these reporters had also caught the spirit. The spirit of empirical method and inductive technique was present to be sure, but the spirit of half-disciplined delight was often equally strong. Part of the delight was the joy of things, things not as poor shadows of some ideal reality but solidly affecting and significant in themselves. The nature reporters felt the exhilaration of cosmic justification as they scrutinized the intricacies of hornets' nests, watersheds, eclipses. It seemed they touched the reality by collecting its specimens and charting its fluctuations. Both the tone of their reportage and contents of their biographies illustrate this joyful motivation.

The natural philosophy community valued, as the science community still does, a plain style, rigorous logic and persistent skepticism. It disavowed appeals to pathos and other forms of subrational or prerational speech. But the reporters themselves responded to complex emotions, and were, after all, in part products of the old cosmos of culture. Reading their letters, diaries and narratives—even essays and logs—one discovers in their work incompletely suppressed delight and love for all the details they observed and collected as they pursued reality in nature. Their joy, their feeling of justification and their passionate advocacy suggest a parareligious motivation and content to their work. And properly so, for these naturalists were apostles of reality and evangelists of the good news that things signify truth, contain beauty and justify all energy devoted to their description and collection. Reality may be difficult to capture and define intellectually, but emotionally it is powerfully magnetic. And reality was what the new science was all about. It is what great cultural movements are usually all about —Christianity, Communism, secularism. The pursuit of reality is an urgent, joyful and humanly transforming activity. And those who numbered the streaks of the tulip were thus energized, delighted and transformed by their mission and witness.

With the consent of their patrons and public, the naturalists brought a new world into existence. They transformed the cosmos by making the nature of the new world and of the ancient East familiar to all who could read or be read to. They sought and found novel remedies for old and new diseases, through the practices, herbs and folk customs of foreign places. They exploited new geographical perspectives to establish the parallax with which to compute distances in the skies and cultural distinctions among races. They dug up fossils that challenged accepted beliefs about the age and history of the world, and they described animals and plants not accounted for in the bestiaries and herbals of antiquity—questioning thereby the accepted theories of creation. And they lent a new dimension to romance by grafting geographical mobility and new-found Edens to the stock of old myths of the fall, lost innocence, lost tribes, pilgrimage and exodus. As a group they contributed to the cluster of forces—economic, political, religious, intellectual and social—that forever transformed the plausibility structures of western culture. Individually they coped with culture and nature, inventing strategies, deploying fictions, accommodating paradox. Their human doing and being is what speaks to us across time, not their science as science or sensibility as norm.

II

The Nature Reporter in
Colonial American Culture

Culture and Nature in America

Broadly speaking, culture is humankind's characteristic strategy to "fix" itself and the world it inhabits—"fix," in both senses, meaning to set and to remedy and improve. Such culture work is what distinguishes humankind from other animals. Nonhuman animals live in a closed world not of their own making; humans inhabit an open world which they articulate and define by their own symboling and patterning. The geosphere and biosphere exhibit patterns too, of course, which impinge on the lives of both humans and animals, but it is the "homisphere" of our own and of our predecessors' making that specially controls and informs human existence—our perception of ourselves and our kind and our relation to the rest of the world. Devised patterns, arbitrary attributions and consensual perceptions bind the artificial world which blankets the given one, interacts with it and comes itself to seem given and objective. Cultures define the cosmos, but not for all time and not neutrally.[1]

The second entry in "The Diary of Adam and Eve," by Mark Twain, shows the pair beginning to fix the world by naming its givens:

Tuesday Been examining the great waterfall. . . . The new creature calls it Niagara Falls—why, I am sure I do not know. Says it *looks* like Niagara Falls. That is not a reason, it is mere waywardness and imbecility. . . . The new creature names everything that comes along, before I can get in a protest. And always that same pretext is offered —it *looks* like the thing. There is the dodo, for instance. Says the moment one looks at it one sees at a glance that it "looks like a dodo." . . . It looks no more like a dodo than I do.

In the same precritical way that Eve knows a dodo by the look of it, persons brought up in western culture know nature when they see it. And culture. Nature and its opposite, culture, are equally categorical cultural fictions, and one of the persistent problems for members of western cultures has been to decide which phenomena ought plausibly to be sorted into each. There is slight difficulty in assigning cathedrals to culture and foxes to nature, but certain phenomena, especially when they are new or alien to the original culture, create a kind of steady dissonance. Aboriginal Americans are an example. First they were "Indians," a familiar type of alien person within the cognitive map of western culture. They were later designated "savages," "devils," "animals," degenerated Jews or protohumans by those who sorted them first into piles labeled "nature" and "culture" and then assigned them proper stations within these larger categories. The problem is that cultural categories are never simple, but belong to a complex of other categories that define how one is to feel, act toward or perceive objects of one category as opposed to another: one acts differently in a confrontation with a "noble savage" than with a "devil."

For European philosophers the problem was academically challenging; for American settlers, vital. And that is one of the chief distinguishing characteristics of American as opposed to "visiting" naturalists coping with American nature. Sometimes naming mattered little. If you call a thrush a robin it never protests. On the other hand, if you call a bison a buffalo, you presuppose its potential for domestication. Geography and astronomy would seem to be minimally charged with cultural resonance but the opposite is true. To chart and locate watersheds radiates intimations of centering the cosmos in America; to take a parallax on the transit of Venus from Philadelphia serves culturally to "establish" Philadelphia in the cosmos. Even the simple discovery of a new plant like the Venus's-flytrap or a new animal like the opossum declares the specialness of America, thus challenging the uniquely central position of Europe. In the politics of cosmos establishment and maintenance, American nature became a test and a challenge to the presumptions of conventional culture. It became finally a weapon in the arsenal of American establishment.

To cope somehow with nature is essential to every culture.

Some cultures are so heavily traditional as to seem nearly static. Others are eager for change, regarding change within certain limits as an evidence of continuity. This has long been the case with western culture. The conquest and organization of nature at home or abroad has been one of its chief preoccupations— and triumphs. American culture shares with European culture this dedication to the "nature project," but a nature project in service to an imperial culture differs markedly from one developed at its perimeter. At home in England a nature reporter needed only to accommodate a familiar nature, one already categorized, its items exploited or ignored according to ancient custom. The facts of weather, vegetables and minerals already belonged to the culture in some fashion and it remained only for the new philosophers to alter explanation patterns, demythologize mistletoe, freshly apprehend seashells or native bats. Domestic culture overlaid familiar nature congruently, if not homologically. Exotic nature imported in fact or by report could be assembled into categories and distributed at some leisure.

At first American nature seemed alien, exotic, foreign to European culture. Habits of handling it sometimes failed to work. Consider the rattlesnake. Even weather seemed untamed. Old myths of deluge and exodus took considerable stretching to embrace the fossils and tribes of the new world. Recipe knowledge about herbs and game, commonplaces about travelling, all became problematic, victims of nature shock. But the earliest reporters abroad in America persisted and fed the stuff of America back to the center where it could be absorbed or adapted, where it sometimes resisted "acculturation" and stimulated new assessments of the nature of nature. These reporters, even when they lived and worked in America, were "in service" as it were to European culture. Like the honey bees introduced to America by Europeans—and which the Indians called the "whiteman's fly"—reporters scouted for new spots to swarm but returned the essence of American nature to the mother-hive. They produced a wild honey and wax that sweetened and enlightened European life. As they danced their triangular dances of intelligence they sometimes conveyed a very un-bee-like sense of their own worth and delight in what they did.

Their story is important, but more interesting yet is the

behavior of certain later, native American reporters—reporters with a strong sense of the legitimacy of American nature in its own particularity and as a base for an American culture. Their work suggests a different metaphor. Like spiders, they threw webs of their own devising across gaps in the cosmos they inhabited, connecting plants to uses, snakes to warnings, meteorological phenomena to portents, themselves individually to the nature of America. The work of these "spiders" was the more difficult and original job, demanding that the individual reporter's self be included. Consequently they supplied richer cultural insights. They worked at the wildest interface of culture and nature, confronting the liveliest puzzles generated by the nature project. Living in nature which sometimes radiated intimations of design not easily accommodated by familiar culture, the American reporters altered myths or invented new ones. They found themselves driven to formulate new recipe knowledge, new symbolic connections between place and meaning, even new myths to legitimate their ethos. They did so and in the process helped to build a fresh and distinctive literature in which the peculiar opportunities and qualities of American nature supported the notion that America contained a unique significance for humanity. They transformed the "nature project" into the "American project." [2]

The Work of the Bees

They were not bees, of course, but persons of complex loyalties and interests. Still on the whole they delivered "exotic" nature to civilization. The earliest reporters did not often distinguish between fancy and fact or science and rhetoric. They wrote whatever they had reason to believe their parent culture would value; fact, hearsay, exaggeration and encomium appear side by side in their work. Attempting to attract settlers to the new world John Smith wove news of nature's plenty into his advertisement of the good life to be had by colonists in America:

Heer nature and liberty affords us that freely, which in *England* we want, or it costeth us dearely. . . . He is a very bad fisher [that] cannot kill in one day with his hooke and line, one, two, or three hundred Cods. . . . If a man worke but three dayes in seaven, he may get more then hee can spend, unless he will be excessive. [*A Description of New England*, 1616] [3]

From Salem in New England Francis Higginson dispatched a description of the country that spoke as strongly to man's dreams of ease and abundance in nature as did Smith's, yet contained hard intelligence as well:

Fowles of the Aire are plentifull here, and of all sorts as we haue in *England* as farre as I can learne, and a great many of strange Fowles which we know not. . . . Here are likewise aboundance of Turkies . . . farre greater then our English Turkies, and exceeding fat, sweet and fleshy, for here they haue aboundance of feeding all yeere long, as Strawberries . . . and all manner of Berries and Fruits. [*New England's Plantation*, 1630] [4]

Not infrequently a reporter's too great admiration of nature's wonders produced reports in which hearsay or fanciful interpretations of natural phenomena distorted the plain truth. Thomas Morton wrote that the beaver's tail has a "masculine vertue for the advancement of Priapus" and, more amazingly, that beavers build square houses on dry land by "pyling one [log] uppon another" and then "move it to a pond." [5] Morton's work is enchanting, littered throughout as it is with his poetry, playful language, and extravagant cussing of the Pilgrims. And he conveys through his narrative a vivid impression of the license and abundance he enjoyed in nature, but very little hard information.

Later reportage is controlled, veracious, plain and objective by comparison. It sometimes sacrifices elegance to objective content, but it compels the reader's attention with fascinating details. Consider John Lawson's description of the bison:

The Buffelo is a wild Beast of America, which has a Bunch on his Back as the Cattle of St. Laurence are said to have. He seldom appears amongst the English inhabitants, his chief Haunt being in the Land of Messiasippi, which is, for the most part, a plain Country; yet I have known some killed on the Hilly Part of Cape-Fair-River, they passing the Ledges of vast Mountains from the said Messiasippi, before they can come near us. I have eaten of their Meat, but do not think it so good as our Beef; yet the younger Calves are cried up for excellent Food, as very likely they may be. It is conjectured that these Buffelos, mixt with our tame Cattle, would much better the

Breed for Largeness and Milk, which seems very probable. Of the
wild Bull's Skin Buff is made. . . . These Monsters are found to weigh
(as I am informed by a Traveler of Credit) from 1600 to 2400 Weight.
[*A New Voyage to Carolina,* 1709] [6]

How different from Morton's descriptions of beasts. Here Lawson consistently tells his reader not only facts about bison, but
whether his information derives from hearsay, judgment, conjecture or his personal investigation in phrases like "are said
to have," "very likely may be," "it is conjectured," "I am informed by a Traveler of Credit" or "I have eaten of their Meat."
Wild beasts from the haunts of "the Land of Messiasippi" are
sufficient to capture the reader's wonder without decorating the
report with extravagant fabrications.

To justify the tone and texture of his prose the eighteenth-
century nature reporter often praised common sense, skepticism,
plain style and fact. He deprecated the mendacity, naïveté and
extravagance of earlier (generally French) reporters. In his Preface to *A New Voyage to Carolina,* John Lawson (fl. 1700–11)
declared, "I have laid down Everything with Impartiality, and
Truth, which is indeed, the Duty of every Author, and Preferable to a smooth Stile, accompanied with Falsities and Hyper-
boles." He further notified the reader that, "I have been very
exact, and for Method's sake arranged each Species under its
distinct and Proper Head." A generation later Cadwallader
Colden (1688–1776) declared his *History of the Five Indian
Nations* (1727, 1747) to be "genuine, and truly related," and
then echoed Lawson by adding, "where Truth only is required;
a rough Stile with it, is preferable to Eloquence without it. . . .
I have sometimes thought that the Histories wrote with all the
Delicacy of a fine Romance, are like French Dishes, more agreeable to the Pallat than the Stomach, and less wholsom than
more common and courser Dyet." [7] Though nature reporters
conventionally avowed such discipline, they often practiced a
greater latitude. Enthusiasm for their subjects drove them to
accept and transmit accounts of rattlesnakes charming their
prey, Indian cures for syphilis, opossums with "nineteen" lives:
"if you break every Bone in their Skin, and mash their Skull
leaving them for Dead, you may come an hour after, and they
will be gone away," Lawson wrote. [8] Here as elsewhere, his delight in things finds expression in his prose, but only a tone-deaf

and singularly humorless reader could be misled by this exaggeration into taking it for literal fact. The nature reporter's subject was nature; his purpose and role, to transmit nature's beauty and worth as well as truth to his readers. For that task he called upon whatever felicities of speech and devices of persuasion he had mastered. A later age would somewhat differently define what was persuasive and acceptable in reportage, but for these prescientific reporters, truth first, but not exclusive of subjective response, was the goal.

Commonly nature reporters communicated the facts they or others had found in travels away from home, and their reportage continued to resemble earlier accounts of discovery and exploration. John and William Bartram travelled north and south on natural history expeditions. Professor John Winthrop of Harvard sailed to Newfoundland (1761) to make astronomical observations of the transit of Venus. Mark Catesby, Alexander Wilson and John James Audubon walked the back country in search of birds. Jonathan Carver, Major Robert Rogers and Lewis and Clark journeyed across the continent to survey its geography, biology and inhabitants. Though any colonial could have dispatched descriptions of backyard nature (and some did), nature as a rule was "out there." [9]

To this day much American nature writing retains this "adventure-in-the-wilderness" and "tour-of-the-prairies" quality. Like their seventeenth- and eighteenth-century predecessors, ninteenth- and twentieth-century nature writers like Henry Thoreau, J. J. Audubon, John Wesley Powell, John Muir, John Burroughs, William H. H. Murray, Florence and Francis Lee Jaques, Edwin Way Teale and Joseph Wood Krutch also frequently wed geographical mobility to the retrieval of natural knowledge. Even "at home" nature seems an alternate reality. Twentieth-century American vacationing non-naturalists manifestly prefer to meet nature away from home—at the seashore, in the mountains or at "camps" in the "wilderness." And in the city nature is often segregated into parks and museums. Living in the country and writing about nature for an urban audience, Hal Borland (nature essayist for the *New York Times* Sunday editorial page) claims to feel "like a foreign correspondent reporting an alien scene." [10]

It is difficult for modern readers to recapture the problematic air of American nature for the early reporters. It was not a matter simply of an unfamiliar plant here, a slightly different snake there. The whole design of American nature seemed unsettling, as if anything might happen when one left civilization. One of the most interesting descriptions in this respect is James Stirling's "An Account of a remarkable Darkness at Detroit, in America" (1763). Stirling reports an occurrence which in earlier times would have provoked prophecies of doom:

I got up at day break: about 10 minutes after I observed it got no lighter than before; the same darkness continued untill 9 o'clock, when it cleared up a little. We then, for the space of about a quarter of an hour, saw the body of the Sun, which appeared as red as blood, and more than three times as large as usual. The air all this time, which was very dense, was of a dirty yellowish green colour. [A breeze from the Southwest finally] brought on some drops of rain or rather Sulphur, and dirt, for it appeared more like the latter than the former, both in smell and quality. [Royal Society of London, *Philosophical Transactions,* hereafter RSPT, 53:63]

Stirling's report is exact in details and expressive of wonder and apprehension. Yet even more interesting is the awesome ignorance about midland geography that emerges as Stirling attempts to explain the cause of the darkness. He says that three theories were entertained. The "vulgar" French and the Indians believed it to be a plague brought by the English at Niagara. Others thought it was due to the burning of the woods. But Stirling, doubtless with a natural philosopher's interest in recent eruptions of Vesuvius, thought "it most probable, that it might have been occasioned by the eruption of some volcano, or subterraneous fire," which sent "sulphurous matter" into the air to mix "with some watery clouds" and fall back down as dirty rain ("Detroit," p. 64). If as late as 1763 educated Englishmen could rationally posit volcanoes in the Ohio-Mississippi basin, there was no limit to speculations about the nature one might find in the interior of the new world.

A similarly significant article by Peter Collinson (1767) reported the discovery by George Croghan, Sir William Johnson's deputy, of "monstrous teeth" and "elephant" tusks "near 7 feet long" on the banks of the Ohio four miles past "Miame" (RSPT 57: 465). Similar remains found in Siberia and elsewhere had interested the Royal Society for years, and Sir Hans Sloane had written an exhaustive commentary on them as early as 1727.

But wonder and perplexity characterize Collinson's speculations because he is unable to account for the skeletons in America by the "universal deluge" theory that had been used to explain their presence in Siberia. Presumably "the Flood" had floated the carcasses to Russia, but he never supposed it had floated them as far as America. Nevertheless the Scotch surgeon, William Hunter, F. R. S., decided that the teeth from Ohio were those of a meat-eating mammoth. "Though we may as philosophers regret it," he said, "as men we cannot but thank Heaven that its whole generation is probably extinct" (RSPT 58: 45). Note the significance of *probably*. The interior of America was unknown, and the natural philosophers had over the years been forced to adopt such a mixture of credulity and skepticism that they were no longer able to scoff at the idea that volcanoes and carnivorous mammoths existed beyond the frontier. No wonder, then, that nature reporters who travelled to the interior of America were unable to suppress emotional excitement at the arcana they discovered, and no wonder they reported even the tallest tales told by Indians and trappers, as well they should have.

Yet once the existence of credulity and awe at the very heart of eighteenth-century *ethnoscience* has been noted, an important qualification must be added.[11] Even in reporting great novelties these reporters exhibited a rational control of credulity and a predisposition to interpret even odd phenomena in accordance with established truths. Remarkable darkness and suns as "red as blood" are not interpreted as signs of God's supernatural interference in the world, but are related to the eruptions of Vesuvius of which all Georgian virtuosi were aware. In fact, it may have been partly such cautiousness that prevented the natural philosophers from apprehending the truly revolutionary implications of the teeth from Ohio and carcasses in Siberia. Nothing in their cognitive maps prepared them for the inference that a half-million and not a half-dozen millenia had passed since life first appeared on earth. Nothing in their skepticism, sharpened on countless outrageous travellers' lies, encouraged their acceptance of prehistoric monsters significantly different from animal species that moved through the nature they catalogued.

Next to "old-fashioned" science, the quality of nature reportage that most startles new readers is the reporters' personal presence in the report. They address the reader directly, speculate freely about the import of what they find, indulge in digressions and even humor. They group anecdotes, tall tales, particular observations, personal narratives, speculation, moral indignation and humor together with Indian glossaries, maps and catalogues of plants and animals as if all were equally acceptable forms by which to convey what they had gathered. In this they demonstrate the unsettled state of nature study in the eighteenth century. No standard content, form or rhetoric had been established beyond a certain preference for plain style, general honesty and (in biology and physics) binomial denomination or mathematical "fluxions."

Yet the language of nature reportage was the language of the educated general public, and its structures were the conventional ones of literature or the oral tradition. Professor Winthrop wrote simple, lucid, direct and plain expository prose; Paul Dudley, urbane and articulate descriptions; Conrad Weiser, terse, abbreviated journals; Jonathan Carver, travels which included extravagant anecdotes; and William Bartram, narratives in which lyric effusions compete with Latin catalogues of plants. Such latitude allowed and perhaps fostered reports that displayed a personal flavor, vernacular spice and enchanting richness of content surprising to those acquainted only with modern scientific literature.

Reporters often occupied the center of the world they described, with the first person singular pronoun "I" as an emblem of their own and humanity's centrality. Young Jonathan Edwards's report on the "flying spider" beautifully illustrates the balance between subject and object that occurs in many nature reports:

I have severall times seen in a very Calm and serene Day at that time of year, standing behind some Opake body that shall Just hide the Disk of the sun and keep of his Dazling rays from my eye and looking close by the side of it, multitudes of little shining webbs and Glistening Strings of a Great Length and at such a height as that one would think they were tack'd to the Sky by one end were it not that they were moving and floating. . . .[12]

Writing in 1755 to Peter Collinson in England, Benjamin Franklin reported exactly and economically his investigation of a whirlwind he and his son had encountered, but he described not only the whirlwind's behavior but his own as well:

> I followed it, riding close by its side, and observed its licking up, in its progress, all the dust that was under its smaller part. As it is a common opinion that a shot, fired through a water-spout, will break it, I tried to break this little whirlwind, by striking my whip frequently through it, but without any effect.... I accompanied it about three quarters of a mile, till some limbs of dead trees, broken off by the whirl, flying about and falling near me, made me more apprehensive of danger.... When we rejoined the company, they were admiring the vast height of the leaves now brought by the common wind, over our heads.[13]

Franklin's personality, knowledge, physical energy and inclination to test theories combine to make this piece the kind of satisfying human report that it would not have been otherwise.

Of necessity, the earliest nature reports by Americans were avocational productions of those whose primary duties lay in managing plantations, governing colonists, surveying the realm, ministering to the inhabitants—aboriginal or pioneer—or simply surviving in the new habitat. They reported when the press of other business was slack, or when something interesting, useful or odd occurred to force the occasion. Thus plagues, earthquakes, mammoth bones, medicinal herbs and meteorological disturbances constitute a large portion of the avocational reportage of men like Cotton Mather, William Wood, William Penn, Roger Williams and William Byrd II. Few of them defined their niche in culture as chiefly that of nature reporter.

In the eighteenth century nature reportage became a distinct vocation for some. Men like Mark Catesby, John Bartram and Jonathan Carver did much more than dispatch occasional and rare intelligence of the new world. They strove for accurate and comprehensive intelligence of nature, and reported usual as well as arcane phenomena, digested as well as raw data. Unlike avocational reporters they devoted long years and sustained effort to the description of nature and conceived of themselves as having found a calling against which other activities took second place. In all cases they deliberately acquired new skills (sometimes late in life) so that they could pursue their vocation competently. Carver learned surveying; Bartram, Latin; and

Catesby, engraving. Each kept abreast, as well as he could, of what other reporters were doing and thinking.

Vocation is not profession, yet the distinction is one of degree. Many of the accoutrements of professional discipline had begun to influence eighteenth-century nature studies—special tools, vocabularies, techniques and paradigmatic models of reality. One thinks primarily of the Linnaean system in biology and the Newtonian in physics. Editorial control to some extent regulated reports submitted to society publications. Critical reviews of published work increasingly appeared in official transactions, lay magazines and interpersonal correspondence like that between John Bartram and Peter Collinson, Alexander Garden and Linnaeus. Financial and other patronage controlled certain projects, especially those requiring expensive equipment like telescopes, costly logistical support by ships, or defrayment of publication cost. In the early years of the century the British government or private patronage supplied most of the support to reporters and explorers. As the century ended, the popularity of natural history and its inclusion in the curriculae of academies and colleges created a market for nature magazines. Charles Willson Peale's museum, John Bartram's arboretum, and many other gardens, collections and libraries became repositories of natural data and centers of study for naturalists.[14]

Something akin to supernatural support also encouraged reporters to persevere in their vocation. Legend has it that John Bartram found his calling as a botanist suddenly one day while plowing a field. Looking at a daisy he was shamed to realize that he had destroyed many flowers without having learned their structures and uses. This "inspiration suddenly awakened" him and converted him to botanical studies, we are told. This account has been attacked as the fiction of J. Hector St. John de Crèvecoeur, the "American Farmer." But his fiction is plausible in general even if too simple and romantic. Bartram's culture presupposed the power of inspiration and the reality of conversion, and had legitimized several options for expression of such energy—religion, botany, *belles lettres*. Bartram chose to be a botanist, and his choice satisfied him and his culture. His energy found a channel alternate to traditional Christianity or humanism. He could not have persisted had all his neighbors, his wife and himself categorized botanical pursuits as mad or evil. He required at least minimal cultural tolerance to succeed.[15]

Actually, of course, Bartram and others found more than tolerance for their vocation. From the earliest times explorers had been directed to answer questions about native populations, indigenous natural productions and passages to India and elsewhere. With the establishment of natural philosophy, this query-response pattern became rationalized. Formal queries informed reporters of stories worth investigating and of "beats" to be covered. Jonathan Edwards described the flying spider because an Englishman who corresponded with his father had asked for intelligence of new-world spiders. Also behind that query were Martin Lister's earlier questions about spiders and the nature of their silk. Similarly, "Directions for Sea-men, bound for far Voyages" (1662–63) showed sea captains which meteorological, cartographic, astronomic and biological data to recover during voyages, just as Henry Oldenburg's queries (1672) directed those bound for "Hudson's Bay" to retieve all sorts of data on the flora, fauna, natives and climate of that part of the world. Thomas Jefferson's *Notes on the State of Virginia* (1784) answered queries directed to him by the Secretary of the French Legation in Philadelphia. Queries travelled in both directions across the Atlantic and up and down through the colonies. Men like Samuel Williams and Benjamin Franklin wrote letters to fellow colonists in an effort to discover the drift of weather across the land. Alexander Garden, John Clayton and others often queried European virtuosi—Linnaeus, Gronovius, Dillenius—for information about discoveries and for guidance in understanding and classifying specimens. In a sense, even odd specimens sent from America to England were mute queries to virtuosi. Implicit and explicit queries exercised a pervasive control upon the practice of nature reportage in America.[16]

What ought to be termed "tacit queries" exercised an equally pervasive control over curiosity, perception and reportage. Internalized from long familiarity, these curiosities almost invariably led authors to include certain features in their reports. Nearly every substantive book of nature reportage includes a section on Indian manners, customs, religion and language. To suppose that each author had been personally asked to supply such material, needlessly complicates the issue. Each simply knew and felt that a continuing thirst for intelligence of that type justified whatever efforts might be made in that direction. Similarly, the ubiquity of essays on how to exploit natural products, even in works where they seem out of place, signifies the widespread cultural assumption that such knowledge was integral to understanding and appreciating nature. Thus Mark Catesby's *The Natural History of Carolina,* primarily a book of information about flora and fauna, contains a brief account of how to make caviar and a longer one on "The Manner of Making *Tar* and *Pitch.*"

In certain instances inherited presumptions generated the structure of nature reports. Raymond Phineas Stearns reveals in *Science in the British Colonies of America* (Urbana, Ill., 1970) the way in which the ancient division of the world's elements into earth, air, fire and water continued to tyrannize the categorical structure of early natural histories. More narrowly, Aristotle's belief that swallows hibernate under water rather than migrate in winter provoked numerous close observations of bird behavior throughout the eighteenth century. And the Paracelsian "doctrine of signatures" furnished hoary support for the tendency of reporters to perceive a correspondence between herbal specifics and particular human diseases: if God had shaped the walnut meat as a visible reminder of its efficacy for brain disorders, logic suggested that He probably also planted anti-veninous herbs in the neighborhood of rattlesnakes. While the list of inherited, perhaps even archetypal, presumptions of western man could be extended indefinitely, it is sufficient here to note that more than nature itself and conscious purpose influenced the American nature reporter.

The journeymen of the nature project were sustained in their calling by diverse social rewards and personal satisfactions. More than thirty colonial residents gained Fellowship in the Royal Society of London, among them Zabdiel Boylston, William Brattle, Paul Dudley, Cotton Mather and Professor Winthrop of Massachusetts; John Winthrop, Jr. and his grandson John Winthrop of Connecticut; Benjamin Franklin of Pennsylvania; Alexander Garden of South Carolina; and John Mitchell of Virginia. Others gained fellowship in additional societies: Garden in the Edinburgh Society, The London Royal Society of Arts and the Swedish Royal Academy; John Bartram in the Swedish Society; and Franklin in the Académie Royale des Sciences. With the institution of the American Philosophical Society even more found recognition of their talents and contributions.[17]

Some found themselves "a species of eternity" in Linnaean denominations—*Gardenia, Claytonia, Catesbaea, Franklinia, Mitchella, Banisteria.* Many more achieved recognition in the pages of the *Philosophical Transactions* of the Royal Society, or other publications which carried their letters, notes, essays and catalogues. Others found their discoveries and observations alluded to, reported or otherwise acknowledged in print by others. A few found the worth of their study "recognized" when it appeared without quotation marks in another man's work. The almanacs, magazines and newspapers of the day featured reprints of material first presented elsewhere. Many, of course, published their works separately.[18]

Certain personal satisfactions also sustained them. Over and over again they reveal in the tone and content of their prose their delight at being among the knowledgeable and initiated. For example, nearly all American travel accounts contain Indian glossaries. By any modern anthropological or linguistic standards these are ridiculously naive and inadequate representations of Indian pronunciation and vocabulary. While they may have served some slight utility in their day, they seem mainly to signify the author's pride in knowing what few others knew. Similarly, travel reports generally included catalogues of plants and animals, usually too vague and random for systematic use. These lists reveal the author's delight in particular skunks, fruits and places discovered in his travels. Featured always in such lists are recipes for the use of novel products—uses discovered by the author himself or learned from natives he met: plants to cure illness, fruits to steep in brandy, ways to find honey, when to eat roast bat. Such items demonstrate the reporter's *naturalization,* his adept, special knowledge of that unfamiliar world.

The humor of these *cognoscenti* reveals links with nineteenth-century western humor—humor at the expense of the greenhorn. It displays the same pride in skill, confidence in practicality and disdain of superstitious illusion that characterizes American humor from Benjamin Franklin to Mark Twain, and beyond. One of the more lengthy pieces of such humor is contained in "A method found out in New England for discovering where the bees hive in the woods, in order to get their honey," published in the 1721 *Philosophical Transactions* (RSPT 31: 148–50). "Although often attributed to Paul Dudley," says

Stearns (p. 459), "the manuscript sources suggest" that this piece "originated" with Paul's father, Joseph. Dudley describes precisely how to capture bees on a plate of honey and release them at widely separate locations so as to determine by triangulation the location of their honey tree. At the end of this exposition, he recounts the experience of an "acquaintance":

An ingenious Man of my Acquaintance the last Year took two or three of his Neighbours that knew nothing of the matter, and after he had taken his Bees, set the Courses the first and second Bee steered, made the off-set, and taken the Distance from the two Stations to the Intersection, he gave orders to cut down such a Tree, pointing to it; the Labourers smiled, and were confident there was no Honey there, for they could not perceive the Tree to be hollow, or to have any hole for the Bees to enter by, and would have disswaded the Gentleman from felling the Tree, but he insisted on it, and offered to lay them any Wager that the Hive was there, and so it proved to the great Surprise of the Country-men.

Country boob and Connecticut Yankee are the personae, confident audacity the gambit, and special but practical knowledge the tool of this parable, the tenor of which is the virtue of initiates and the unregenerate ignorance of outsiders. One is reminded of Thoreau's delight in amazing his "countrymen" by finding arrowheads beneath their feet and of Mark Twain's reports of the relish river pilots took in confounding passengers with apparently inexplicable skills and memory feats.

Some colonial Americans did make significant contributions to the "advancement of science." Stearns argues that "the refined (and somewhat unrealistic?) standards of . . . 'pure' science" by which Franklin alone is admitted to the ranks of science are overly severe (pp. 678–80). More realistic standards allow others to be included in the list of important American scientists, he says: Cotton Mather for "pioneering observations" in plant hybridization, Zabdiel Boylston for his "use of medical statistics to justify" smallpox inoculation, Paul Dudley for his work in hybridization, Isaac Greenwood for an "ingenious" proposal to devise ocean charts, Thomas Godfrey for his quadrant, and James Logan for demonstrating the "sexual reproduction of maise." Stearns cites Dr. Garden's and John Clayton's contributions to Linnaean biology, Ebenezer Kinnersley's refinements of Franklin's electrical work, Dr. John Lining's and Dr. Lionel Chalmers's studies of the relationship between sickness and weather, and Professor Winthrop's "pioneer studies with regard to the mass and density of comets and his early perception of the undulatory character of earthquakes." To this list should be added Mark Catesby for his contributions to the theory of bird migration.

The nature reporter was more than a scientist. He legitimated and domesticated as well as described American nature. Take cultural significance as the criterion instead of scientific worth, and a new picture emerges. Hundreds, not tens, of Americans retrieved and organized data, digested experience, devised technical innovations, published their thoughts, persuaded others what to feel, and taught thousands what to value in America's physical and biological environment. Every magazine or newspaper correspondent who reported some natural fact, even if the fact was neither new nor useful, nurtured the increasingly common assumption that the objects of the natural world had a beauty and worth independent of their valuation by theological or humanistic constructs of reality. Though science had to struggle to survive in colonial America, nature reporting thrived. Nature reportage contained much that was essential to colonial culture—ideas, feelings and values as well as facts. To read it only for historical or scientific information is to miss much of what it has to offer. One must weigh its emotional as well as intellectual content and scrutinize its symbolic as well as logical systems of meaning.

The Intersecting Webs of Culture, Nature and Personality

Volume One of the American Philosophical Society *Transactions* (1769–71, hereafter cited as APST) assembles documents explicitly intended to be useful and shows the mixture of styles, content, tone and format characteristic of such collections. Introduced by an "Advertisement" of eleven pages and a "Preface" of nineteen, its essays and notes cover a variety of topics from the nature of comets to the virtues of mineral springs, from meteorological speculations to opium cures for lockjaw, from Samuel Bard's definitive description of diphtheria to descrip-

tions of "native SILK WORMS" by Moses Bartram. The articles
are accompanied by illustrations, diagrams, graphs and tables.
It seems formally chaotic, like a virtuoso's cabinet, yet when
viewed as a monument of ethnoscience it holds together as a
kind of symbol: an ethnosemantic chart of the culture work
required of American naturalists. The articles mend and tighten
the connection between (1) agriculture and homiculture, (2) the
human and cosmic machine, (3) celestial location and human
place, and (4) the American project and the progress of Man.[19]

One of the most interesting documents in this regard is "An
ESSAY on the cultivation of the Vine, and the making and pre-
serving of Wine, suited to the different Climates in *North
America*" by "the Hon. EDWARD ANTILL, Esq; of *New Jersey*."
Antill (1701–70) was a respected member of colonial society, a
zealot for agricultural improvement, a brewer, an orchardman,
a merchant and a farmer. Though winner of "the largest individual premium ever awarded" to an American by the British
Society of Arts for his contribution to American viticulture, and
author of "the best article on the subject to come out of the
colonies," according to Brooke Hindle, he is mentioned in
neither the *Dictionary of American Biography* nor the *Dictionary of National Biography* and has been generally ignored by
scholars.[20]

Though primarily a detailed and explicit eighty-page man-
ual for beginning vignerons, specifying the planting, fertilizing,
cutting and staking operations to be followed each of the four
years required to establish a vineyard, Antill's essay is prefaced
by three pages of introductory apology, advertisement, promises
of "plainness and honest simplicity" and exhortations to action.
Antill believed it America's destiny to "give the finishing
stroke" of perfection to "an undertaking as antient as the days
of Noah." He saw America as the place for this perfection to
occur, because here we are "free from the force of all [Europe's]
prejudices and erroneous customs ... and [may] judge for ourselves" with the use of reason. The introduction ends, as works
in the genre often do, with an anti-French sentiment: "But let
not the people of America be dupes to France or any set of
designing men." Apparently viticultural improvement was to
Antill a metaphorical equivalent of cultural progress. "An
Essay" appears to be the viticultural correlative of homicultural

tracts like the *Federalist Papers* (1787–88) and *Letters from a Farmer in Pennsylvania* (1767–68).

From Antill's classical and biblical allusions it is clear that vineyards and wine are symbols of the good, settled, productive, orderly society—and that to plant and care for vineyards is to provide for the future. The social, political and ethical tenor of his vision emerges when Antill discusses the necessity of discipline for vines: "it teaches the Vines to become reconciled to a low and humble state, it curbs their pride and ambition, which is always to climb and mount up above every thing that is near them, and educates them to bear fruit within your reach" (APST 1: 140). Elsewhere he says, "a well regulated Vineyard resembles a fine regiment under proper and exact discipline" (APST 1: 134). His ideal is pure, but his practice is temperate: "The greatest difficulty, as experienced Vignerons know, is so to manage a Vine, as to keep her within the height and compass of a frame, and yet to cause her to bear fruit plentifully" (APST 1: 153). Benevolent monarchy, not tyranny or license, is where he puts his faith. So did many colonists in 1769, men passionately American but alert to the potential for chaos they felt lay in natural liberty, uncivilized nature. He says of native American vines that they are "much more intractable than those of Europe [and] will undergo a hard struggle indeed, before they will submit to a low and humble state, a state of abject slavery" (APST 1: 191). His diction ("abject slavery") suggests a covert admiration of colonial intractability, but like so many Americans of the 1760s and 1770s, he was torn between order and freedom as competing norms. He perceived no natural synthesis of freedom and order in wild nature the way some poets and philosophers had begun to do. Wild birds and wasps were pests in the vineyard. He recounts with pride how he shot birds and employed country boys to destroy their nests and eggs to protect his grapes.

Another interesting dimension of Antill's essay is the symbolic identity of human and plant life its diction betrays. In numbering the streaks, not of the tulip but of the grape vine, he created descriptions which are both botanically exact and poetically suggestive. He consistently personified, speaking of buds as "eyes," sap as "blood," cuts as "wounds," stock vines as "mother plants" and offshoots as "nephews," "suckers" and "robbers."

Even though such personification is conventional in viticultural literature, the literal and figurative contents of Antill's essay occasionally coalesce into startlingly vivid images. For example, the unfortunate result of too early pruning, we are told, is that the "parent suffers by her fresh wounds . . . and her eyes are drowned in her own blood" (APST 1: 129). Such examples make one wish that an Edward Taylor, a Philip Freneau or a Frederick Goddard Tuckerman had exploited in a sustained creation the potential for poetry in the facts and implications of viticulture. That no poet adopted viticulture as a vehicle of social, ethical or political thought and that Antill only tentatively indulged in such wit, may be an index of the growing split in eighteenth-century culture; ancient and modern were too far separated to be synthesized into art by any but the most exceptional artist.

Astronomy suggests a different tenor of connection between humanity and nature—a rarified and impersonal one. Nearly half of Volume One of the *Transactions* is devoted to the transit of Venus and related matters. The tone is serious, the style, plain and the contents, instructive. Many reports, short and long, simply repeat and confirm each other, and considerable space is given to graphs and tables. These transit documents are about the most objective reportage imaginable, yet even in their descriptions of equipment, location, experimental controls and views through a telescope, delight and wonder appear. As David Rittenhouse wrote, "Imagination cannot form any thing more beautifully serene and quiet, than was the air during the whole time [of the transit]; nor did I ever see the Sun's limb more perfectly defined, or more free from any tremulous motion; to which his great altitude undoubtedly contributed much" (APST 1: 27). The expression of sublime quiet is not scientifically irrelevant; he reports a natural fact that supports the validity of his measurements. But its greater relevance is cultural and personal, an esthetic fusion of astronomy and celebration. Astronomical precision and feelings of sublimity seem to meet as nature and Rittenhouse mutually hold their breath. James Pearson, in a separate report of the precise moment at which the body of Venus joined the sun, similarly turns to poetic speech: "I saw a continued thread of light, like a silver lace, but still with a tremulous motion" (APST 1: 50).

Another section of the *Transactions* is devoted to reports of ingenious mechanical contrivances: a machine for cutting the grooves in files so automatically that "a blind man might cut a file" with it; a device powered by a ship's progress for pumping bilge out of vessels at sea; a "Self-Moving or Sentinel Register" to control the flue of a chimney and regulate the heat. These inventions admirably illustrate the Yankee ingenuity that many have praised and others ridiculed as a trait of the American character.

One contrivance, however, rises above mere utility and radiates intimations of homology between divine and human mechanics, but less as a proof of God, as with Aquinas's watch, than as a triumph of *Homo faber,* apprentice to *Deo faber.* The lead article in the *Transactions* (placed at the head of the section on astronomy rather than in the section for ingenious mechanical inventions) describes David Rittenhouse's "new Orrery." An orrery is a mechanical planetarium of the sort still used as a visual aid in beginning physics classes. Named "orrery" by Steele to honor Charles Boyle, Earl of Orrery and patron to George Graham, the inventor, this device was no longer novel in 1769. What was notable about Rittenhouse's orrery was its precision and accuracy. Its elliptical orbits were exact and its relative planetary positions accurate "for any time over a five thousand-year period with a margin of error of less than one degree." It was "regarded as symbolic of the precision both of God's universe and of man's science," according to Hindle, and was celebrated in the press as a work of genius that "would have done Honour to the Age and Country in which Newton lived." It may be difficult for twentieth-century persons to perceive orreries as anything but clever visual aids "to instruct 'the ignorant' " and "astonish the skillful," but the energy devoted to its construction and the extravagant praise for the finished product suggest that Rittenhouse's orrery was a kind of icon, an affecting presence, venerable in itself and a careful symbol of the spirit and fact of cosmic design. Confirming that the orrery was an emotional as well as an intellectual construct is the note at the end of the article in the *Transactions:* "N.B. The above machine is to be supported by a mahogany case, adorned with foliage, and some of the best enrichments of scultpure. . . . the clock part of it may be contrived to play a great variety of Music." [21]

One of the boldest documents and one seemingly misplaced in a repository of "Useful Knowledge" is Hugh Williamson's "An *Essay* on the Use of *Comets,* and an Account of their *Luminous Appearance*; together with some Conjectures concerning the Origin of *Heat*" (APST 1: App. 27–36). Williamson's essay is a marvel of daring deduction, entertaining rhetoric, deistic wit and sublime sentiment. He begins by demolishing popular fallacies: that comets burn and were created to house sinners. He ridicules the vanity of supposing that "fifty or an hundred worlds were created for the sake of punishing the inhabitants of this little globe." He also dismisses both Newton's notion that comets "nourish and refresh" the earth's atmosphere and Newton's explanations of the behavior and origin of comets' tails.

In the course of developing his own theory, Williamson announces that heat and light are distinct, that heat is movement of particles but light comes from the sun, that light from the sun has weight and travels with "amazing velocity" ("thirteen millions of miles in a minute") and "considerable momentum" which excites the particles of the comets' tails and "drives" the tails to one side. Light, in short, has mass. He would appear to be verbally, if not mathematically, close to modern ideas in physics, but he arrived there not by induction but by logical inference. Such are the virtues of the grasping intellect. He also deduced, however, that comets are "doubtless" inhabited, and he arrived there by the same route that convinced him it was silly to think that comets were repositories of sinners; namely, it is vanity to assume that the earth alone supports life. Within this one essay, then, Williamson demonstrates both the sublimity and fallibility of rational deduction as an approach to natural truth. But there is no escaping his impulse to assemble a new human and natural universe out of the physical and spiritual shards of the older sacred cosmos.

Extreme particularity and spiritual intoxication exist side by side in the productions of American virtuosi. Particular truths and leaps between the particular Thing and the cosmic Import engaged American nature reporters. What did not much attract their interests was the kind of plodding, continuous, responsible investigation and construction of tentative hypotheses that characterize the work of those engaged in what Thomas Kuhn calls "normal science." Except for the articles on the transit of Venus, there is little evidence in the *Transactions* of that so-called "scientific method" that features the cautious induction of tentative truths from massive data.

The most startling instance of this preference for leaping from minute and useful facts to grand generalizations occurs in the "Preface" to the *Transactions,* exemplifying the similarities of mind between these promoters of "Useful Knowledge" and such enthusiastic explorers as Jonathan Carver. Geographically and climatologically America is likened to China. Protonational enthusiasm is revealed in the surprising hope that, if we manage well, "*America* might in time become as populous as *China*" —an awesome dream. But one discovers that the projected operationalization of this dream is to grow tea and silk in America. While tea and silk were not then irrelevant to empire, the leap implicit in this juxtaposition of Chinese magnitude and plans to exploit "native *Silk Worms* of North-America" (Moses Bartram) is the equivalent in natural philosophy of a mystic's or romantic's perception of the cosmos. Like Whitman's leap from railroads and telegraph cables to "more than India" or Edwards's extrapolation from flying spiders to God's mind, it betrays a kind of willful and drastic fusion of instance and principle, a thirst for signs of divine favor or incarnation in nature. We have come to expect such leaps from Puritans and Transcendentalists, but the avowed rationalism of the Enlightenment has obscured for many the continuity of this tradition in American culture.

The distinctive American flavor of some colonial reporters' work derives in part from their interweaving protonational projects and the nature project. Another source of the impress of culture on science flows from the reporter's lone mediation between self and other, between his personality and the shape and drift of objective nature and culture. The subjective constellation of any one member of culture contains a more or less integrated maze of culture and nature codes. In a time of transition especially, both culturally sanctioned and idiosyncratic customs of apportioning affect *a* to event *x,* meaning *b* to pattern *y,* become freshly problematic. The individual may resolve such matters in many ways: deliberately or automatically, timidly or boldly, perversely or predictably. To the extent that he precipitates evidence of this coping in prose or artifact, he deposits intelligence for anyone who would subsequently chart the probable patterns of invention and accommodation deemed satisfactory and effective by one "informant" from the culture. Extended case studies provide the best model for such analyses (hence the following chapters on Carver, Bartram, Catesby). But an abbreviated and introductory probe will serve to emphasize several modes in the process of culture-nature coping which repeatedly occur in the work of nature reporters. These patterns—of decorum, rhetoric, feeling and knowing—reoccur later in the coping of poets, painters, scientists and others who pay attention to nature and themselves. What humor, play, wit is permissible when dealing with nature? What diction, imagery, grammar or form? What is loathsome or awesome, amusing or neutral? And what does nature mean?

Any number of reporters could serve well as informants. Raymond P. Stearns, Brooke Hindle, Whitfield J. Bell, Jr., and others have identified numerous important but neglected contributors to science; yet collections of primary sources, even the *Philosophical Transactions* of the Royal Society of London, remain to be scrutinized systematically for contributions by Americans. That organ was the primary periodical repository of American nature reportage before 1769. International or non-national in tone, these pieces exude the Americans' sense of brotherhood with their English and continental fellows. John Winthrop of Harvard, for example, contributed "Cogitata de Cometis" in Latin, the *lingua franca* of natural philosophy.[22]

Of the many and varied contributions by Americans, very few have survived the indifference that our culture in general and science in particular feel toward such outmoded technical documents. Historians of science and historians of ideas have developed an interest in certain pieces, especially those that economically illustrate distinguishable patterns of eighteenth-century thought or the origins of modern scientific disciplines. Only a handful of these have attracted wider public interest; one thinks most readily of Franklin's pieces on electricity and kites that often appear in anthologies of colonial American literature. Other contributions to the Royal Society have lain relatively undisturbed and been ignored by students of culture and historians of science. Not all merit resurrection, but those of Paul Dudley, F.R.S., most certainly do. Grandson of

Governor Thomas Dudley of Massachusetts, son of Governor Joseph Dudley, and himself Chief Justice and Attorney General of Massachusetts, Paul Dudley (1675–1750/51) is not unknown to American historians. The *Dictionary of American Biography* calls him an "accomplished naturalist," citing his twelve contributions to the *Philosophical Transactions,* but neither *The Cambridge History of American Literature* nor the *Literary History of the United States* (hereafter cited as LHUS) mentions him at all. Even Moses Coit Tyler expresses no praise of Dudley's literary ability; yet Dudley's twelve essays are among the most delightful and instructive reading in the *Philosophical Transactions.*[23]

The breadth, variety and specificity of Dudley's interests as a naturalist are suggested by the subjects he essayed. He recounts New England methods for making maple sugar, making "molosses" from apples and curing illnesses by sweating. He describes the rattlesnake, the moose, poison sumac, "degenerated" smelt and a "stone" removed from a horse. He reports and theorizes about earthquakes, essays a description of American trees and plants, and describes the types, uses and habits of whales. The essays are remarkable for their precise language, accurate detail, vivid images and sensitivity. Particularly felicitous is Dudley's description of the moose, "a Creature, not only proper, but it is thought peculiar, to North America, and one of the noblest Creatures of the Forest" (RSPT 31: 165). He includes a physical description as any natural philosopher would, but beyond that, he describes the moose so well that the reader can visualize it alive and moving:

When a Moose goes through a Thicket, or under Boughs of Trees, he lays his Horns back on his Neck, not only that he may make his way the easier, but to cover his Body from the Bruise, or Scratch, of the Wood. . . . A Moose does not spring, or rise in going, as an ordinary Deer, but shoves along side-ways, throwing out the Feet, much like a Horse in a rocking pace. [Pp. 166–67]

No one who has seen a moose hurrying out of a bog or trotting along with its chin up can fail to perceive the remarkable accuracy of Dudley's description.

"Shoves along side-ways" is just one example of Dudley's selective use of the vernacular, a practice this graduate of Harvard and the Inner Temple shared with reporters of modest education like Carver and Bartram. He calls naturally upon common images, conveying the strength and size of moose by saying "they will bend down a Tree as big as a Man's Leg" to browse upon it. He mixes usage to describe the "Swamp Sumach": "The inside of the Wood is yellow and very full of juice, as glutinous as Honey or Turpentine; the Wood itself has a very Strong unsavory Smell, but the Juice stinks as bad as Carrion" (RSPT 31: 145). Vernacular and latinate diction, homely and elevated figures, bland and pungent expressions—Dudley takes what he needs to achieve the best report possible.

Dudley's style enriches especially his reports of awful events. He frames his account of an earthquake in Massachusetts by reporting the weather conditions preceding it: "the three last Days of [the month were] so violent hot, that there was no working or travelling by Day, or sleeping by Night" (RSPT 39: 65). His description of the quake itself is an effective combination of subjective experience and objective speculation:

For my own Part, being perfectly awake, though in Bed, I thought at first my Servants, who lodged in a Garret over my Chamber, were ha[u]ling along a Trundle-bed: But, in truth, the Noise that accompanies an Earthquake seems to be *sonus sui generis,* and there is no describing it. . . . My House . . . seemed to be squeezed or press'd up together, as though an hundred Screws had been at work to throw it down; and shook not only every thing in the House, particularly the Bed under me, but the Building itself, and every Part of it so violently for the Time, that I was truly in great Fear it would have tumbled down, and my Family perished in the Ruin. . . . [RSPT 39: 68]

He mistakes the earthquake's rumble at first for sounds of domestic activity, thereby conveying both his own perceptual confusion and the terribly unusual nature of the experience. In succeeding pages, he vividly particularizes the threat of the quake by cataloguing its effect on dishes, chimneys and stone walls. Yet personal as was his reaction to the quake, he supplements his own experience with allusions to Pliny and Aristotle. In the kind of afterthought so common in nature reportage (content spilling beyond form), Dudley appends some observations out of proper sequence: "I shall throw them into a Postscript," he says, with amusing nonchalance and honesty.

Above all else, reporters gather and transmit data. Some

do more. Dudley's way of making a moose present or an earth-quake upsetting introduces us to ourselves as well as to objective nature. We learn how we feel in the presence of such phenomena. He "initiates" us, so to speak, in field experience. His treatment of that culturally resonant curiosity, the rattlesnake, is an especially rich example. His opening paragraph both describes the snake and establishes its unique danger to man:

The Rattlesnake is reckoned by the *Ab-origines,* to be the most terrible of all Snakes, and the Master of the Serpent-kind; that which causes their Terror, without doubt, is their mortal Venom, and the Ensign of it is their Rattle; and it is most certain, that both Men and Beasts are more afraid of them, than of other Snakes; and while the common snake avoids a Man, this will never turn out of the Way. [RSPT 32: 292]

Dudley solidly plants his rattlesnake in our path and displays its fangs and rattle. He limns the rattle with careful precision in a later paragraph, doing verbally what Catesby later would do graphically (see fig. 15): "Our People . . . soon discovered . . . the Tail to be compos'd of Joints, that lap over one another, somewhat like a Lobster's Tail; and the striking them one upon another, forms that Noise, which is so terrible to Man and Beast" (p. 293). Also like Catesby, Dudley is sensitive and meticulous in his attention to color and minute details. The "three sorts" of rattlesnake are "distinguished by their Colour, *viz.* a yellowish Green, a deep Ash Colour, and a black Sattin" (p. 292). One especially large specimen was found to have "between seventy and eighty Rattles, with a sprinkling of grey Hairs, like Bristles, over his Body" (p. 295). This truly is a snake of vivid and particular presence.

Dudley makes his snake more threatening by linking it to comfortable domestic images, calling its color "black Sattin," its inactivity "sleepy," and by likening its unique tail (the "Ensign" of its venom) to the familiar and beneficial "Lobster's Tail." Furthermore, he emphasizes the snake's singular and terrible stare ("one is apt, almost, to think they are possest by some Demon") by describing its effect on harmless inhabitants of the woods:

I dare not answer for the Truth of every Story I have heard, of their charming, or Power of Fascination; but yet I am abundantly satisfied from many Witnesses, both *English* and *Indian,* that a Rattlesnake will charm both Squirrels and Birds from a Tree into his Mouth. A Man of undoubted Probity . . . told me . . . he observ'd a Squirrel in great Distress, dancing from one Bough to another, and making a lamentable Noise, till at last he came down the Tree, and ran behind a Log: The Person going to see what was become of him, spied a great Snake, that had swallow'd him. [P. 293]

Here he transmits even hearsay if it works to convey the snake's effect on man and beast.

Generally a more skeptical witness, Dudley often corrects popular misconceptions. He denies a common belief still entertained today that snakes "traject" their venom or leap more than their length when they strike (pp. 294, 292). Momentarily reassuring as such denials may be, he never underplays the snake's mortal danger. He reports that the rattlesnake's celebrated warning rattle, loud and clear in dry weather, does not occur in wet, "for which Reason, the *Indians* don't care to travel in the Woods, in a Time of Rain, for fear of being among these Snakes before they are aware" (p. 293). And later he describes a snake as "full five Foot and an half long, and as big as the Calf of a Man's Leg" (p. 295). Given the usual location of snake bites, this juxtaposition of snake and "Calf of . . . Leg" chillingly conveys the danger one feels from such snakes.

Formally, this piece oscillates between cool description and moments of alarm. The rhythms, in short, complement Dudley's emotional vacillation as he contemplates this ambiguous symbol of nature whose coiling and striking is quick but whose slowness in ordinary motion is "an Instance of the Goodness of God" (p. 292). Perhaps Dudley's response to such ambiguity is implied in his choice of an ending for this account, a report of the annual spring slaughter of the snakes: "They generally den among the Rocks in great Numbers together; the Time of the retiring is about the middle of *September,* and they don't come abroad till the middle of *May,* when our Hunters watch them, as they come out a sunning, and kill them by hundreds" (p. 295). Perhaps some vital catharsis results from slaughter of teleologically ambiguous species. At any rate Americans still practice this sport.

Knowing how to take nature, how to infer its message or purpose, is the final and most difficult task generated by the nature project. Dudley's article on whales and whale fishery

(RSPT 33: 256–69) is fascinating in this regard. Filled with tenderness, awe, reverence for life, mercantile rationality, violence in nature and systematic human butchery, Dudley's piece repeatedly reminds one of Melville's classic attempt to master rationally and symbolically his perception of the significance of whales and whaling. Dudley begins in rationality. He admits that much has already been said of whales, announces his "endeavor to avoid Repetition," and proceeds to classify the several whales—the "Right," the "Scrag Whale," the "Finback," the "Bunch or humbpack Whale," and, climactically, the "Sperma Ceti," which he thinks should be called the "Ambergris Whale." Following this four-page cetological catalogue, Dudley launches into six pages of remarks on the reproduction of whales, their domestic economy and behavior, and man's "Way and Manner of killing Whales," and ends with a description of natural "Enemies of the Whale."

Having been lulled into detached objectivity by the tone of the opening section, the reader is startled by the description of the domestic life that follows. We are told that cows so cherish their young that they will not abandon them, even in danger, and that should a calf be harpooned the cow, "so soon as ever the Calf is dead . . . grows so violent, that there is no managing her" (p. 261). Their "Manner of the Propagation" anticipates Melville's celebration of "dalliance and delight" in the depths of the sea:

When the Cow takes Bull, she throws herself upon her Back, sinking her Tail, and so the Bull slides up, and, when the Bull is slid up, she clasps him with her Fins. A Whale's Pissel is six Feet long, and at the Root is seven or eight Inches Diameter, and tapers away till it comes to about an Inch Diameter: his Stones would fill half a Barrel, but his Genitals are not open or visible, like those of a true Bull. [P. 260]

One hardly knows how to sort out the messages in this description. The mating itself sounds remarkably tender and is easily accommodated within anthropomorphic models, but the gigantic scale of the bull's genitalia is outside all terrestrial experience and expectation. The casual juxtaposition of the two elements derails the imagination conditioned by literary conventions. Ordinarily such discord ought to trigger mirth or horror, but neither emerges here. Sympathy and objectivity co-here (or at least coexist) more easily in nature reportage than in *belles lettres*.

The kind of healthy "wholeness" apparent in Dudley's de-scription of whale coitus yields to an appalling combination of cool rationality and mercenary interest when he describes the operations required to secure ambergris from whales. Sig-nificantly, Dudley seems not at all disturbed by the methods employed to extract from a *dead* whale a product useful to man:

Mr. *Atkin's* Method of getting the Ambergris out of the Whale was thus; after the Fish is killed, he turns the Belly upwards, and fixes a Tackle to the *Penis*, then cuts a Hole round the Root of the *Penis*, through the Rim of the Belly, till he comes to the Entrails, and then searching for the Duct or Canal at the further End of the Bag, he ties it pretty near to the Bag, and cuts the Duct off beyond it, upon which he draws forth the *Penis* by the Tail, and the Ambergris Bag, entirely follows it, and comes clean and whole out of the Belly. [Pp. 268–69]

That is the entire paragraph. It is as plain and cool and exact as a description of how to make sap into maple sugar. Sensitive to the emotional resonance of phenomena, Dudley is also capa-ble of discriminating between what seem to him valid and in-valid inferences of symbolic content. Here he displays no more squeamishness about dead whales than Huck Finn does about dead cats.

Comparisons to *Moby-Dick* suggest themselves often in Dudley's piece. One in particular is especially significant for students of American culture. It is a scene of dramatic and savage violence in nature that outdoes even Melville's descrip-tion of similar occurrences in *Moby-Dick* and *Mardi*. Though Dudley nowhere articulates the horrific cosmic implications of the slaughter he describes, he renders the scene vividly. And since it follows the sections describing the filial tenderness of whales, it can hardly be read without an emotional reaction:

The Enemies of the Whale, . . . Killers[,] are from twenty to thirty Feet long, and have Teeth in both Jaws that lock one within an-other. . . . They go in Company by Dozens, and set upon a young Whale, and will bait him like so many Bull-dogs; some will lay hold of his Tail to keep him from threshing, while others lay hold of his

Head, and bite and thresh him, till the poor Creature, being thus heated, lolls out his Tongue, and then some of the Killers catch hold of his Lips, and if possible of his Tongue; and after they have killed him, they chiefly feed upon the Tongue and Head.... The Killers are sometimes taken, and make good Oil, but have no Whale-bone. The Carcases, or Bodies of dead Whales in the Sea, serve for Food for Gulls, ... as well as Sharks, for they are not very nice. [Pp. 264–65] [24]

What one notices in Dudley's amazing evocation of violence, aside from the ghastly particularity of the whales catching hold of the lips and even the tongue of the "poor Creature," is the absence of any sort of deduction about cosmic indifference to suffering. He does not, as Melville does in his poem, "The Maldive Shark," for instance, comment upon the slaughter as "horrible"; nor does he extract implications derogatory to God's goodness as does Queequeg at the end of "The Shark Massacre": "de god wat made shark must be one dam Ingin." Dudley is clearly aware of the cruelty, as his epithet "poor Creature" demonstrates, but his sympathy for the creature does not infect his perspective, which is that of a rationalist reporting the habits and uses of whales. He concludes his remarks with an appraisal of the killer's utility for humans: they "make good Oil, but have no Whale-bone."

Perhaps Dudley's prose suggests one cause for the neglect of nature reportage by many historians. It is more responsive to the demand's of man's emotional nature than is consistent with modern scientific and historical standards of excellence, more sympathetic to technical exploitation of nature than Romantic and post-Romantic literary sensibilities can tolerate, more full of a sense of assurance that man occupies the central link in the chain of being than is palatable to philosophically and theologically alienated persons of modern times. What are we to make then of such antediluvian work? [25]

We can begin by responding as fellow humans to works in which we find "after all, a place for the genuine. / Hands that can grasp, eyes / that can dilate, hair that can rise / if it must" (Marianne Moore, "Poetry"). Their meaning may not be our meaning exactly, for they operated out of a somewhat different ethnosemantic context. While we share certain institutional patterns with them as well as metapatterns of language and cognition, still the exact psychological meaning of their expressions remains a matter of inference—careful and controlled inference, but still inference. And value is even more difficult to capture than meaning. Beauty for them may be ugliness for us; their cosmic truth our laughable superstition. Modern studies have so relativized value and meaning that consent to past arrangements and even faith in our own seems vain and foolish at times.

Peter Berger in his fine little book, *A Rumor of Angels*, suggests a strategy for theologians caught in the same dilemma. In a brilliant stroke he relativizes the relativizers (himself included) and returns to "prototypical human gestures" for assurance that everyman's *ordering, hoping, damning, play* and *humor* are "signals of transcendence." It is the *making* that *means*. And matters. Here is a path the student of culture can follow. What is important and useful about the nature reporters and their work is not their dated formulations but the facts of how they worked. They teach us about acting human, about coping and inventing in a world partly familiar and partly disturbingly unconditioned. And they teach us better because their plausibility structures differ from ours. What seems self-evident to us was problematic in that world; what was tentative then has worked out for us in ways they never anticipated. The dynamics of culture and personality, of being human, is their legacy; the fragility of invention and the resilient intractability of conditionality, their warning. I like their self-reliance and energy, their mythologizing and demythologizing, their doubting and celebrating. For their shortsightedness and too easy assurance that the rest of history would turn out as they presumed, I feel compassion not disdain. Their flaws are their flaws; ours, ours. They did not make our world, neither can they save it. Whatever attenuated psychological influence they still exert to push western civilization into ever more arrogant assaults on nature shrinks to insignificance beside the massive thrust of economic institutions. The nature reporters' potency lies elsewhere, in their demonstration of how to survive and thrive within, even deflect somewhat, whichever culture project is in fashion.

The three men to whom the remaining chapters are devoted offer case studies, as it were, of such surviving and thriving in the face of new realities and ongoing cultural conditions.

Each came from a natural and cultural environment important in colonial life: Jonathan Carver, from New England; John Bartram, from Pennsylvania; Mark Catesby, from England. Taken together they touched all the main regions and ecosystems open to Americans, and they tried all the media and genres of reportage open to naturalists. Individually, they worked out of who they were and what they experienced and worked up their own strategies for making sense of the world and sharing their knowledge with others. Though long gone in one sense, they live on in their works and present compelling instances of human being in early America.

III

Jonathan Carver,
A Connecticut Yankee

Good News from the Interior Parts of North America

In the twentieth century nature reportage survives as a marginal and distinctly minor genre. For the colonial period it was the fastest game in town, just as ecclesiastical polity had been the fever of an earlier time and political or industrial change the excitements later. The nature reporters were ingenious improvisers upon cultural formulations; their readers, the most avid audience an author could desire. The public devoured natural histories, gobbling them whole, consulting them piecemeal for particular intelligence, cannibalizing them for reprint in magazines or repose in private journals, plagiarizing them, digesting them and reincorporating the news they contained into new compendiums of natural knowledge. One type of reportage for which readers especially thirsted was the travel narrative in which they found both original news and standard intelligence. Jonathan Carver's *Travels Through the Interior Parts of North America, in the Years 1766, 1767, 1768* (1778, hereafter cited as T) is a felicitous and well-developed specimen of this type, earning "a greater reputation than any other book of American authorship in the eighteenth century," according to Henry Nash Smith (LHUS, 760). Before the century was out, public demand caused printers in such widely scattered cities as Reutlinger, Edinburgh, Boston, Leyden, Philadelphia, Yverdon, Paris, Dublin and Hamburg to issue a total of fifteen editions of the *Travels*. By 1881 the number has increased to thirty-eight, with Stockholm, Charleston (S. C.), Glasgow, Walpole (N. H.), Braunsweig, New York, Tours and Galizao (Greece) joining the list.[1]

Impressive as this popularity is, the book's real interest for
students of American culture lies below the surface in its "politics" of language, genre, culture and experience—politics in
the sense that what is at stake in the text, in its publication, in
the author's treatment of style and form, is a perpetual negotiation between what are felt to be contending forces. Some are
cosmic: nature and civilization. Some are social, as between
colonists and England or one hustler and the establishment.
Others are intrasystemic, as between genre and genius or facti-
city and significance. Carver negotiates the mazeways presented
him by his culture in order to report truly both what the world
wanted to know and what personal excitement he recalled having experienced in his journey into the continent. Politics is
the art of generating and organizing power—power for personal
and public projects. The projects Carver pursued might be
termed, in descending order of magnitude, the culture project,
the nature project, the American project, and the Carver project.
For each, material and energy had to be transformed into power;
by naming, explication, rhetorical strategies and tactics—old
arrangements disestablished and new configurations legitimated
between thing and significance, place and meaning, person and
culture. Carver's uneven but often startling successes (and, of
course, failures) manifest objectively these deeper dynamics, and
his choices illuminate the metaprescriptions behind his choices.

At the surface, however, is where we must begin and where
Carver's readers began—at the explicit, the avowed, at the communication interface of author and reader. Behind the reader
lay strata of expectation and habit ready to be tapped; behind
the author lay raw experience and patterns of coping with it
and transcribing it to prose. The written result is news: news
of rivers and waterfalls, beasts, savages, useful productions of
nature; news of possible connections between civilized man and
the uncivil world; news perhaps even of hidden designs, desti-
nies to be manifest, alternate worlds to be settled. Profit and
delight is what the reader expected; to instruct and please, the
author's aim. "The adventurer upon unknown coasts, and the
describer of distant regions, is always welcomed as a man who
has laboured for the pleasure of others, and who is able to en-
large our knowledge and rectify our opinions," Samuel Johnson
explained, but he that undertakes to enlighten and please "must

offer new images to his reader, and enable him to form tacit comparisons of his own state with that of others" (*Idler* 97).

New images are precisely what Carver delivered: images of snakes, wild rice, clear waters, deep caverns, Indian heroines and braves. But he also wrote of ancient ruins, lost civilizations and geographical patterns too grand to be spoken of calmly, for these images fit into older myth patterns and linked those myths to a new place, a new time. They made the interior regions of North America a new arena for ancient contests and put an individual man at the center of the new cosmos. Little wonder that his images captured the attention of such men of letters as Coleridge, Wordsworth, Chateaubriand, Schiller and William Cullen Bryant. Little wonder either that later explorers searching for what Carver had found either criticized him when they failed to recapture his vision of reality, or borrowed his words to express their own ineffable feelings in the wilderness.

With the opening sentence of the *Travels* the reader is jerked away from all that is familiar and is transported with Carver "to Michillimackinac; a Fort situated between the Lakes Huron and Michigan, and distant from Boston 1300 miles." Now Boston is a city all can understand, a city like all cities, special in certain ways but essentially ordinary, with shops and churches, civil arrangements, continuity, order. "The Interior Parts of North America" were something else: a land of forts with exotic names at places called Green Bay and Detroit where "Hurons, Miamies, Chipéways, Ottowaws, Potowattimies, Mississuages" and others followed leaders like the famous Pontiac. Out there lived Indians—like Castañeda's Don Juan—with special powers to call rattlesnakes (T, 43–45) and foretell future events (T, 124–29). In such a setting even the simplest phenomena become luminescent. Things signify; even the plainest, most matter-of-fact reports of flora and fauna seem to radiate intimations of deeper meaning. Yet the matter-of-fact remains at the foreground of the reportage.

In the more than five hundred pages of the *Travels,* Carver conveys the news to the reader of novel and delightful, useful and terrible American flora and fauna: news of the wood duck "perching on the branches of trees, which no other kind of water fowl . . . is known to do"; of the "excellent sport" of chasing loons, "exceedingly nimble and expert at diving"; of

"Wood Pigeons" in "such prodigious quantities" that they "sometimes darken the sun for several minutes"; of the "King Bird" that is able to "bring down a hawk," and of the "whipper-will," whose return in the spring tells the Indians that "the frost is entirely gone" and that it is time for them to "sow their corn." He communicates the sagacity and utility of the beaver and the "great delicacy" of the "rich and luscious dish," moose lip and "Carrabou" tongue. He describes the horrors of the "Carcajou" (wolverine), who is a "terrible enemy" to the caribou, moose, elk and deer, and springs down on his victim from trees, brings him to earth by encircling him with his tail, and kills him by fastening upon his neck "and opening the jugular vein." He describes the rattlesnake and rattlesnake plantain, an "efficacious" herb in treating rattlesnake bite that flourishes precisely during the season when "the bite of these creatures is most venomous." From him the reader learns of wild rice that grows eight feet tall and is "the most valuable of all the spontaneous productions" of the interior and may "in future periods . . . be of great service to . . . infant colonies"; of "Ginsang, . . . considered as a panacea" in the Orient; of crab apple trees; of maple trees that produce sugar. At his best, Carver combines accurate and precise description of natural objects with a sense of their significance beyond mere utility and a delight in their being.[2]

Intelligence such as this is the backbone of reportage—specific and concrete reports of nature's plenty, variety, harmony and utility. At its best this reportage provides images that not only delight and instruct but that also encourage the reader to "form tacit comparisons of his own state with that of others" who like Carver, are free to roam through nature with civilized sensibilities, or like the Indians, are free to pursue their lives in nature as children of the forest. But how does a Yankee reporter act in such a setting? In such a way as to carry his rationality into the scene to elevate his own sense of superior competence, and at the same time, flatter the sophistication of his readers. Carver describes a spring thunder storm which terrified the Indians with its violence: "The peals of thunder were so loud that they shook the earth; and the lightning flashed along the ground in streams of sulphur; so that the Indian chiefs themselves, although their courage in war is usually invincible, could not help trembling at the horrid combustion" (I, 86). Yet unlike the savages, Carver himself, though "never more affected in [his] life," wisely avoided running under the trees with the Indians, "chusing rather to be exposed to the peltings of the storm than to receive a fatal stroke . . . that might ensue from standing near any thing which could serve for a conductor." He knew his Franklin and enjoyed applying his knowledge. The account makes clear that for him the most satisfying result of his rational response to natural terror was that "the Indians were greatly surprized, and drew conclusions from it not unfavourable to the opinion they already entertained of [his] resolution" (T, 85).

At other times Carver shows a simple credulity, as when he transmits information gathered from the Indians about the lands to the west and makes what must be one of the understatements of the century. Speaking of the "four most capital rivers on the Continent of North America, viz. the St. Lawrence, the Mississippi, the River Bourbon, and the Oregon or the River of the West," he says, "the waters of the three former are within thirty miles of each other [essentially correct]; the latter, however, is rather farther west" (T, 76). It is actually 700 to 1,000 miles further west, depending upon what is taken as the head of the Oregon. Similarly, reports of trainable rattlesnakes (T, 43–45), Indian clairvoyance (T, 123–29) and herbal cures for rattlesnake bite (T, 482–83) indicate a suspended skepticism that may result from a wish to keep an open mind, to be "free," as he says, "from every trace of skeptical obstinacy" as well as from "enthusiastic credulity" (T, "An Address . . .").

Carver's rationalism in no way mutes his enjoyment of the bizarre, the incredible, the inexplicable. In this he exemplifies the balance between skepticism and wonder that can be seen even in the *Philosophical Transactions* of the Royal Society of London, where theories that swallows hibernate through the winter in mud at the bottoms of streams shared equal status, apparently, with Mark Catesby's rational and persuasive argument in "Of Birds of Passage" that they migrate south. This balance can be seen, as well, in the *Transactions* of the American Philosophical Society where Dr. Hugh Williamson's argument that comets are habitable coexists with Rittenhouse's careful reports of his astronomical measurements and Dr. Samuel Bard's classic description of diphtheria. Even the humor Carver frequently displays in the *Travels* has its counterpart in the works of other natural philosophers of undisputed worth—John Bartram, Paul Dudley, Mark Catesby and of course, Benjamin Franklin.[3] Eighteenth-century readers could identify with Carver. So, apparently, could late-nineteenth-century Americans. More than a century after publication of the *Travels*, Moses Coit Tyler concluded: "Besides its worth for instruction, is its worth for delight; we have no other 'Indian book' more captivating than this" (*The Literary History of the American Revolution*, hereafter cited as LHAR, 1:149–50).

What captivates is not so much the news itself—that instructs and delights—but the opportunity Carver offers the reader to use the book as a gateway to the interior through which he may then travel vicariously, becoming initiated with Carver, naturalized to a region where the mythical and mundane overlap, where fact is pregnant with intimations of design. Carver is especially good at devising scenes that suspend him (and the reader) at the point of overlap between profane and sacred, dangerous and safe, horrid and humorous.

One learns from Carver how to travel through nature—down rivers, across inland seas, up rivers, across portages and down rivers again, as in the Michillimackinac-Green Bay-Fox River-"Ouisconsin" River route to "La Prairie La Chien" on the Mississippi. And the reader learns even how to feel while canoeing on Lake Superior, an experience that suspends one between earth and sky, promotes links between fact and fancy, and rewards scrutiny of concrete objects with emotional intoxication:

The water in general appeared to lie on a bed of rocks. When it was calm, and the sun shone bright, I could sit in my canoe, where the depth was upwards of six fathoms, and plainly see huge piles of stone at the bottom, of different shapes, some of which appeared as if they were hewn. The water at this time was as pure and transparent as air; and my canoe seemed as if it hung suspended in the element. It was impossible to look attentively through this limpid medium at the rocks below, without finding, before many minutes were elapsed, your head swim, and your eyes no longer able to behold the dazzling scene. [T, 132–33]

Anyone who has canoed on Lake Superior will appreciate the force and accuracy of Carver's account. Indeed, William H. Keating, the talented, otherwise scrupulous author of the early-nineteenth-century *Narrative of an Expedition to the Source of the St. Peter's River* and frequent belittler of Carver's accuracy in the *Travels,* was apparently so impressed by Carver's description that he "borrowed" all its essentials for his own later account:

> ... pebbles can be distinctly seen at a depth of more than twenty feet. The canoe frequently appears as if suspended in air, so transparent is the liquid upon which it floats; the spectator, who remains too long intently gazing at the bottom, feels his head grow giddy, as if he were looking into a deep abyss. [2: 199]

Equally forceful in its ability to capture and hold the reader's interest is Carver's description of the "Wakonteebe" cave in the sandstone bluffs of the Mississippi in what is now St. Paul, Minnesota; but it suspends the persona not so much between earth and sky as between the present and a mysteriously pervasive antiquity. Again Carver's participation in the scene is crucial. The cave—"Carver's Cave," as it has been known ever since—extends an "unsearchable distance" back over transparent water, he tells us. Its walls are covered with ancient "Indian hieroglyphicks" cut into the "stone so extremely soft that it might be easily penetrated with a knife." When he "threw a pebble towards the interior parts of it with [his] utmost strength, ... it caused an astonishing and horrible noise that reverberated through all those gloomy regions" (T, 63–64).

Elsewhere Carver invests mundane geography with cosmic significance, suggesting by his tone and diction the existence of realms of reality beyond rational understanding. His report of the four-river watershed, "not ... paralleled on the other three quarters of the globe" (T, 77), suggests a buried metaphorical significance for the interior of North America: "And a river went out of Eden to water the garden; and from thence it was parted, and became into four heads" (Gen. 2:10). One finds as well covert intention to discover that there had been "giants in the earth" (Gen. 6:4) in Carver's reports of ancient fortifications below Lake Pepin, of mountains in the river and of rocks like "ruinous towers" (T, 55–59).

No giants remained in the land Carver travelled, but Indians did, and their sensibilities, morals, habits and histories more than once lent support to his sense of the decorum appropriate to life in nature. A young "prince" of the Winnebagoes "amazed and charmed me" with his "artless yet engaging manners," Carver says. This noble savage addressed the "Great Spirit" at the Falls of St. Anthony and as an offering, "first threw his pipe into the stream; then the roll that contained his tobacco; after these, the bracelets he wore on his arms and wrists; next an ornament that encircled his neck, composed of beads and wires; and at last the ear-rings from his ears; in short, he presented to his god every part of his dress that was valuable: during this he frequently smote his breast with great violence, threw his arms about, and appeared to be much agitated" (T, 67–68).

Less noble but equally engaging were the "robber bands" of Indians along the banks of the Mississippi who unsuccessfully ambushed Carver's party (T, 51–53). Nature does not guarantee nobility any more than civilization does. The "Chipeways" residing near the head of the "Chipeway River" in what is now Wisconsin "seemed to be the nastiest people" he "had ever been among": "I observed that the women and children indulged themselves in a custom, which though common, in some degree, throughout every Indian nation, appears to be, according to our ideas, of the most nauseous and indelicate nature; that of searching each other's head, and eating the prey caught therein" (T, 104).

How titillating to the Georgian audiences of Boston and London. In two instances especially does Carver form portraits of Indians that convey admiration and wonder. Decidedly bawdy in flavor, they demonstrate the relish with which Carver viewed Indian cultures. The anecdotes also illustrate his and his culture's presumptions about Indian liberty and license. But best of all they show Carver adapting eighteenth-century narrative conventions to the work of nature reportage.

In the first tale Carver recounts what an Indian had told him of an escape from captivity in the Wisconsin woods sixty years earlier. In it he exhibits the compressed form, rapid pace and playful tone necessary to the successful anecdotalist, reflecting his mastery of what must have been a thriving oral tradition at all frontier fur factories. According to the tale, French missionaries and traders had sent a party of French and Indians up the Fox River from Green Bay to surprise their enemies in reprisal for many "insults" received in the past. They succeeded and were returning to Green Bay with their captives when "one of the Indian chiefs. . . , who had a considerable band of prisoners under his care, stopped to drink at a brook" while his companions went on. His resulting vulnerability provided one of the captive women the opportunity to free herself and save her people by a novel but direct act:

She suddenly seized him with both her hands, whilst he stooped to drink, by an exquisitely susceptible part, and held him fast till he expired on the spot. As the chief, from the extreme torture he suffered, was unable to call out to his friends, or to give any alarm, they passed on without knowing what had happened; and the woman having cut the bands of those of her fellow prisoners who were in the rear, with them made her escape. [T, 40–41]

"This heroine," we are told by Carver, "was ever after treated by her nation as their deliverer, and made a chiefess in her own right, with liberty to entail the same honour on her descendants: an unusual distinction, and permitted only on extraordinary occasions." Since Carver just prior to his travels had lived in Montague on the Connecticut River near where Mary Rowlandson had suffered her captivity during King Philip's War and where her tale and others like it must surely have been part of the local lore, we are justified in viewing Carver's tale as a variant species of the long popular captivity-narrative genre— "variant" because this heroine is an Indian woman. Probably this accounts for the substantial difference between his tale and those told by Thoreau of Hannah Dustan (*A Week*, "Thursday") and by Mrs. Rowlandson of herself; that is, it accounts for Carver's willingness to forego the pious corollaries so long conventional in tales of Caucasian female captives of Indians, and accounts for his obvious relish of the tale's sexual pivot.

In another anecdote, the "Remarkable Story of one of the Naudowessie [i.e., Sioux/Dakota] Women," Carver pens a tale that must surely be counted as an addition to the tall-tale genre of American frontier stories. "Whilst I was among the Naudowessies," he says, "I observed that they paid uncommon respect to one of their women, and found on enquiry that she was intitled to it on account of a transaction, that in Europe would have rendered her infamous." As a young woman "she had given what they termed a rice feast," he was told, "an ancient but almost obsolete custom" which "scarcely once in an age . . . any females are hardy enough to make." Having "invited forty of the principal warriors to her tent" where she stuffed them with rice and venison, she concluded the feast by regaling "each of them with a private desert, behind a screen fixed for this purpose in the inner part of the tent." By this "profusion of courtesy," she won "the favour of her guests, and the approbation of the whole band"; the young men all vied for her hand and the principal chief "took her to wife, over whom she acquired great sway, and from whom she received ever after incessant tokens of respect and love" (T, 245–46). Carver may have been told this tale by the Indians, or he may have picked it up from traders and soldiers in the West. On the face of it the tale is a factual report. On the other hand, its extravagant content, elevated diction and playful circumlocutions should alert us to the likelihood that whatever the factual basis of this story may be, Carver has consciously crafted a humorous rather than a simple and ingenuous report. Especially convincing in this regard is his parenthetical insertion early in the account of an editorial aside in which he applies Hamlet's remarks upon Dansh drinking customs to the rice feast, writing, it is "an ancient but almost obsolete custom (which, as Hamlet says, would have been more honoured in the breach, than the observance)," an outrageous pun in this context. It is clear that Carver has larded the lean meat of his factual reportage with items that are meant to amuse as well as inform.

Now, travellers are notorious liars—even plagiarists—and Carver has been burned on both counts. But if they have duties as reporters, they also have privileges as raconteurs, and one of the privileges is to improve on reality and sometimes even to "put on" the gullible or inattentive reader, especially if (as in this case) the reader is English and the traveller, a native, colonial American. Two items in the first edition struck knee-jerk skeptics as incredible: an account of a Grand Portage Indian "priest" of the Killistinoes who mounted an elaborate hocus-pocus and then prophesied the arrival of a group of traders the next day (T, 123–29); and an anecdote about a Menomonie Indian and a Frenchman. This Indian, who kept a rattlesnake ("his god") in a box, released the snake one October in the presence of the Frenchman and told him that the snake would return to its box in May. The Frenchman was skeptical, the Indian confident, and the result was a wager. In the spring they met in the woods and the Indian hustled the Frenchman into escalating the bet. Then "behold on the second day, about one o'clock, the snake arrived" (T, 43–45). At first glance Carver seems to have been duped by the Frenchman who relayed the tale to him. But he ends the account by saying "the French gentleman vouched for the truth of this story, and from accounts I have often received of the docility of those creatures, I see no reason to doubt his veracity." *Docility* is the tip-off. If there is one attribute that every account of rattlesnakes makes clear has no place in rattlesnake nature, it is docility. But his readers in England were "sold," nevertheless, just as they were to be time and again by the tall tales of Benjamin Franklin, J. J. Audubon and countless other Americans.[4]

Carver added "An Address to the Public" to the second edition of the *Travels* (1779) that is classic in its way. Rather than confess either that he himself had been taken or that he had intentionally put one over on his readers, he compounds the "put-on." Assuming an air of injured reputation and earnest explanation, he endeavors "to eradicate any impressions that might have been made on the minds of his readers, by the apparent improbability of these relations." He declares his own freedom from "every trace of skeptical obstinacy or enthusiastic credulity," hardly a rhetorical coup de grâce since his credulity is at issue. He defends the story of the rattlesnake as having been told him by a Frenchman "of undoubted veracity," but again, if there is one quality all Anglo-American readers had come to regard as problematic in Frenchmen, it was veracity. The general mendacity of French reporters had become an article of common sense among readers of colonial-period travels, and must surely have been a standard ingredient of the frontier oral tradition in which Carver became immersed, just as it had been in the New England he came from. Either Carver was incredibly gullible and dull, or he knew what he was up to. It seems obvious to me that the latter was the case. What Carver does in the "Address" is to offer the readers taken in by the original anecdotes an opportunity to entangle themselves further in error, the opportunity to laugh again at this credulous bumpkin who persists in asserting the plausibility of incredible tales. But it is they whose prejudices and relaxed attention have been "worked" again by the old traveller. Ironically Carver's hoax lived longer than anyone could have imagined. Twentieth-century historians have been caught in the trap, misled perhaps by the portrait of Carver in the third edition which makes him look like a humorless, pompous Englishman (properly the butt of jokes rather than the perpetrator) rather than a Yankee. They should have read more alertly.[5]

One of the more fatuous "truths" of folk wisdom is the simplistic formulation, once a liar, always a liar. S. I. Hayakawa might say that lie_1 is not lie_2 is not lie_3. Lies are inventions, verbal compositions which playfully alter the culturally normal interrelations between mundane actuality and the fictions that allow us to order meaning and value. Grand fictions of this sort which impose improbable meaning on worldly facts are called myths, visions, religion. Smaller lies are called poems, stories, biographies and the like, but in neither case is it common or reasonable to assume that the creators of such fictions are characteristically untrustworthy. Yet, in nature reportage, hoaxes and jokes and lightness of tone are often taken as symptomatic of the general invalidity of the whole work. At least one historian, Edward Gaylord Bourne, made Carver's habit of "borrowing" from other authors into a kind of lying and concluded that the man could not be trusted, concluded even that he may not have written the *Travels* at all. The avowed purpose of reportage—to collect and report what was useful in nature—misled Bourne, as it has others, into demanding that reportage be either so devoid of human voice as to appear objective or that subjective and objective elements be clearly labelled and separated. No sophisticated eighteenth-century reader demanded that, as several later historians pointed out in answer to Bourne. Since then, and since the discovery of Carver's manuscripts in the British Museum, historians have credited Carver's claim to making the journey—but they are often still uncomfortable with his free mixture of fact and fiction, even though the two are quite distinguishable if one attends to genre and tone as well as denotation, to the whole fabric and not just the lexical monofilament.[6]

While Carver's small fictions entertain, his larger ones inspire. Having collected reports from the "Killistinoe" and the "Assinipoil" (i.e., Cree and Assiniboin) Indians of the lands to the west, he digests report and legend alike to create a description with mythic overtones of Paradise in the West. According to the tale (T, 118–23), in the remote West live a "nation" of "smaller and whiter" tribes who cultivate the ground and the arts. They live to the west of the "Shining Mountains" and have "gold so plenty among them that they make their most common utensils of it." Tradition has it that they are descendants of the preconquest Mexican kings and that they "fled from their native country to seek an asylum in these parts," Carver tells us. They live in innocence in the "most interior parts" and shun contact with outsiders, remembering as they do the "monsters vomitting fire," which magically killed harmless Indians at astonishing distances. Among the "Shining Mountains"—which "from an infinite number of chrystal stones, of an amazing size, with which they are covered, and which, when the sun shines full upon them, sparkle so as to be seen at a very great distance"— these simple people "found a place of perfect security." Perhaps some future Columbus or Raleigh may yet discover, concludes Carver, this land

to the west of these mountains . . . where future generations may find asylum, whether driven from their country by the ravages of lawless tyrants, by religious persecutions, or reluctantly leaving it to remedy the inconveniences arising from a superabundant increase of inhabitants; whether, I say, impelled by these, or allured by hopes of commercial advantages, there is little doubt but their expectations will be fully gratified in these rich and unexhausted climes. [T, 122]

Here is the Myth of the Garden, El Dorado, Crèvecoeur's asylum, the Western Reserve and the "safety valve" all rolled into one. Carver must be counted in the ranks of American myth-makers. Or if not quite a myth-*maker*—few men are that—then a "medium" of the emerging and converging myths and symbols which in constellation served to legitimate the American secular project—"secular" rather than "sacred" because, in spite of the inspired language and rhythms, the project itself grew out of the new naturalism and its half-brother, mercantilism. The natural intelligence retrieved and organized in Carver's *Travels* is well sprinkled with remarks on mines, fertile pasturage, navigable waterways, routes to the Indies and Gulf of Mexico, waterfalls, wild game and the like. And the concluding "Appendix" (T, 527–43) is an explicit promotion piece, keyed to the maps in the volume and directing governmental planners to the potentials for development of an inland empire.

Should Carver then be blamed for the excesses of a later age that pursued his vision, settled the interior, exploited nature and "civilized" the Indians? Hardly. The force of myths such as those Carver penned is problematic; they cannot easily be shown themselves to motivate behavior. But they do serve to articulate, sanction and legitimate it. They organize a rhetoric that makes the imperatives acted out by people seem plausible, worthy, admirable and worth sacrificing present pleasure for. And this was surely one of Carver's contributions to the American project. The other was a vision that the nineteenth century largely ignored and which has only in the last half of the twentieth century regained momentum in the ecology movement. He envisioned a world in which self-reliant persons could live in considerable harmony with nature and with others of different cultural traditions. The designs of nature and the designs of humans were for him compatible, providing occasions for delight, awe, playfulness, useful work. He became a secular prophet who spoke to a much wider audience than Crèvecoeur did; to surveyors, promoters, settlers and others who wanted intelligence of the interior. He seemed the ethical proof that ordinary men could move west, parley and live with Indians, and even—to the readers of the third and later editions of the *Travels*—acquire large tracts of land from the Indians for development.[7]

The Individual and Culture

How does a person like Jonathan Carver become a secular prophet? What prepares him, allows him to turn himself into a politician of the cosmos and weld into a plausible coalition the diverse rhetorics of economic, spiritual, national and natural interests? This is a question with two dimensions, one of evidence, one of argument. Unfortunately, as is often the case with nearly anonymous men who achieve fame late in life, much of the documentary residue of Carver's early life is lost, problematic or sketchy. Of course his *Travels* is another sort of evidence just as any work of art is. And the characters of other men of his time and condition also help document the parameters that limited or fostered Carver's development. But the tenuous nature of the evidence in Carver's case only emphasizes what is true for all biographers but is less noticeable when facts abound: one must reconstruct the man, create a plausible model that accounts for all the facts, both *unique* as they emerge from documents and his art, and *common* as they emerge from comparative analysis, for Carver was a type as well as an individual. We all are, and that is the drama of biography.[8]

The eighteenth century was not flooded with nature reporters any more than the nineteenth was flooded with mechanical inventors. Most Americans in both centuries lived lives of "quiet desperation" or quiet accommodation. But Carver and his sort—the nature reporters—bolted the usual and probable for the special and exceptional. They "found themselves" or their callings and often found them late in life rather than early. And from this they drew some of the special flexibility, special marginal quality, that allowed them to do what was unexpected, unschooled, improvisational or inventive. Most people are probably gradually if unevenly socialized and stray only under extreme pressure—religious, economic, military, etc. But those who make a revision late in their lives, convert to a new plan, have forever after a peculiar perspective on their culture and on the solidity of its values and preconceptions. As numerous "personal narratives" of religious conversion make clear, the life drama of those converted contains three phases: (1) a beginning worldly period of common living, with its ups and downs of excitement and disgust; (2) a middle sacred experience, prefigured sporadically, experienced intensely for a short period and evolving into a somewhat longer development of the full meaning of the new life; and (3) a final period lasting until death (or subsequent reconversion) in which the subject copes again with the world of ordinary institutions, conventional expectations, biological and social demands—a period of agony in which the struggle to persist is met by the inertia of others and the intractable elements remaining within oneself from the earlier life. This is the pattern of Carver, a Yankee who found his vocation late in life, embraced it and gave himself up to its logic, allowing it to organize his practical skills as observer and mapper, and to channel his mental and physical energies. From then on he either travelled in search of natural knowledge and adventure or reported what he experienced in the wilderness.

Jonathan Carver was born in Weymouth, Massachusetts, April 13, 1710, of educated and "able stock on both sides." If he was not a descendant of John Carver (first governor of the Plymouth colony) as once was thought, his roots did go back to John's brother, Robert, of Marshfield, Massachusetts, an off-shoot from the Plymouth Separatists. Jonathan Carver's family moved to Canterbury, Connecticut, in 1718 when he was eight years old. He probably studied medicine for a time, but did not become a physician. He married Abigail Robbins in 1746 and "later moved to Montague, Massachusetts, where he was a select-man in 1759." There is evidence that he "made twenty pairs of shoes for Moses Field in 1754," but that he was a shoemaker by trade is not otherwise established. As early as 1746 he may have been a private in the Connecticut forces, and records show that he rose from the rank of sergeant in 1758 to a captaincy of a Massachusetts regiment in 1760, serving "honorably" under General Wolfe at Quebec and General Amherst at Montreal during the French and Indian Wars. It is clear also that he was present at the siege and capitulation of Fort William Henry in 1757 and that his account of it in his *Travels* is "veracious and reliable." [9]

In his endorsement of one of Carver's later petitions for financial relief or remuneration, General Gage attested to Carver's military ability and gave him the "character of a very good man." Evidence, in addition to that supplied by the *Travels,* shows that he was a skilled draughtsman, surveyor and map-maker "well known in geographical circles years before his book made him still better known." In short, from the external and corroboratory evidence a picture emerges of Carver as a Connecticut Yankee, adept at several trades, married and the father of two daughters, educated, skilled in mapmaking, and a soldier who had served his king and his country well in the glorious activity of driving the French from North America.

Thus Carver appears to have been a mildly interesting and worthy man. But nothing in these particulars anticipates the Carver that burst forth in the *Travels*—visionary, hustling, irrepressible, aggressive, self-assured. Even the attestation of General Gage has the air of an automatic letter of recommendation. Carver had made little mark on his society by the age of fifty-three and the few facts we know were mostly resurrected and published only in the twentieth century after attacks on Carver's veracity had stimulated other historians to defend him, prove that he was literate, had really been in the army as he said, had made the travels he claimed. Half a century of life had deposited few objective remains, but we can guess at some of the subjective residue of his experiences. Born during Queen Anne's War, reared during the religious and financial ferment of the early years of the century, married the year after New England troops vanquished the French at Fort Louisbourg, enlisted in the new wars with France in 1758, captured at William Henry and triumphant at Quebec and Montreal, Carver was understandably proud of his service and anxious to continue his adventures in behalf of "King and Country," not less because he found himself suddenly discharged and at leisure by the Peace of Paris. His New England heritage had conditioned him to expect that nature should serve man and that America was the arena in which man's destiny was to be realized. Personal energies had often been channelled into deeds of seeming cosmic import in New England, and Carver naturally imbibed the New England homology of self, America and the cosmos.[10]

Carver had tried several vocations, but none held him like his adventures in the wilderness. Having travelled in Canada and the Lower Great Lakes region, he had met Indians and helped drive out the French. He had seen firsthand the possibilities offered by the interior and the need for accurate maps and for exploration. Excited by what he knew to be important to England and the colonies, he nevertheless found himself again a civilian and unemployed, sharing that "fate with the rest of my fellow officers of the same Province of being Discharg^{d} without any advantages of half Pay nor of any share of Land in the Conquered country, which was granted to the officers on the Establishment with whom we did equal duty through the whole of the Several Campaigns in the . . . war" (J2, 54).[11] Here are the seeds of discontent—the dissonance between his own enlivened sense of opportunities, and disregard by the "Establishment" of the potential for service he knew he contained. He had tasted a wider usefulness than that promised by village life, and he meant to press on:

In this situation I yet wish[d] for an Oppertunity of Serving my King & Country being for some time out of Employ was very disirous for an Oppertunity of being Employ[d] in Some business whereby I might be usefull in making some further discoveries in those new Acquisitions. . . . I privately procured some Books on [Geography] and studiously endeavoured to inform my self in every science Necessary for a compleet draghts man. [J2, 54–55]

The plan he hit on was to search for the fabled Northwest Passage to the Pacific. Others also held that dream, notably Robert Rogers, but the scheme had long excited the English imagination. The significant thing for Carver biography is not the origin of the dream but his daring reach in making it his own. There were dozens of ex-soldiers around better prepared than he, and he must have known it. But his exceptional character began to assert itself, and he made the crucial break with his ordinary past and with the normal expectations of his culture: in wartime, the army; with the end of war, a resumption of ordinary life. War, like pleurisy (a common preliminary to rebirth in conversion narratives), may happen to any man quite apart from his deserts or preparation, but once come, its fever grips the imagination of the predestined. Nauseous recollections of normal life contrast with new and brighter prospects of personal worth, service to fellow men, and testimony to the beneficence and justice of God—or in this case, rational nature. Carver left no contemporary journal account of this "conversion," so the details, the exact sequence and the tone of the experience is supposition. Carver may merely have been restless and taken up exploring without a clear notion of what it signified. But the facts remain that he broke from his old life, left his family behind, and went adventuring in the wilderness. The conversion model makes sense of much that follows in the biography and is plausible, given the culture from which Carver came. But whether he found his new calling before he left or only later in the woods or later yet at Michillimackinac can never be established. Yet it matters very little. He clearly wove into the fabric and design of his finally published *Travels* whatever intimations of significance he had gained.

Converting his plan into action was not as easy as he had imagined. It took money for one thing; naturally he looked to government. "Sometime in the Year of 1765," he wrote, "being in company with a Gentleman ... with whom I had had the Pleasure of serving in the army and was then an half pay officer on the establishment, ... I communicat^d my desires of Aplication to Government for Liberty to go on some discoveries toward the South Sea" (J2, 55). He seemed to like the idea, but before they could settle upon a "Particular Plan of proceeding," Carver "heard that Major [Robert] Rogers ... had laid a Plan before the Board of Trade by which he propos^d to find a passage Through the Continent to the South Sea. . . ." He thought it likely Rogers's plan would succeed and changed his mind about applying, contenting himself to "wait for Maj^r Rogers' Return to America with a determination of Taking the first opertunity of offering [his] service to Joyn in the opperation of the Plans." He spoke to Rogers in Boston in March 1766 and "Entered into an Agreement" with him. There has been some argument among historians about Carver's debt to Rogers—even the novelist Kenneth Roberts entered the fray, making Rogers the hero of his *Northwest Passage* (New York, 1937), Carver the fool. However, no substantial disagreement remains about the basic facts of the arrangement or the expedition that followed.[12]

Carver probably was recommended to Rogers by the former Ranger, James Tute, who was to become first in command of the expedition for which Carver enlisted. Carver was to be mapmaker and third in command. He left Boston on May 20, 1766, and arrived via the Albany-Oswego-Detroit route at Michillimackinac by August 28 and was formally commissioned by Rogers at "eighte Shillings Starling p Day" to map and explore the rivers, bays, harbors and Indian towns between Michillimackinac, Green Bay, the Mississippi at Prairie du Chien and St. Anthony Falls. He and a group of traders left Michillimackinac on September 3 and proceeded by the Green Bay-Fox River-Wisconsin River route to Prairie du Chien. From there he, a Frenchman and an Indian proceeded up the Mississippi to St. Anthony's Falls (November 17), mapping and exploring the country and dealing with the Indians as they went. At St. Anthony the supplies that had been promised him had not arrived, and he wintered with the Sioux on the Minnesota (then St. Peter's) River, though just how far up the river or with whom he stayed has been the subject of controversy. He claims to have gone two hundred miles up the Minnesota and convincingly describes the landscape.

In April 1767 he left the Sioux to return to Prairie du Chien with an interpreter, Mr. Reaume (fourth in command). They next went north by the Chippewa-St. Croix-Goddard (Bois Brule) route to Lake Superior and up to Grand Portage where they arrived in July. They were to have advanced to "Lake Winnepeek" but waited at Grand Portage for the supplies promised by Rogers. The supplies never arrived, so in August the party started back to Michillimackinac following the north and east shores of Lake Superior. Carver wrote on August 27, "Here ends this attempt to find out a Northwest Passage." [13]

And there ended what must be taken as the seminal experience of the new Carver, an experience he would spend the rest of his life writing and talking about—trying at first to turn it into new projects supported by the government, then into a book about his experiences and what he "knew" they signified.

Carver had spent slightly more than a year west of Michillimackinac when he returned there on August 29, 1767. During the winter Robert Rogers was arrested and jailed for treason, and Carver wrote the first and perhaps second draft of the manuscript "Journals." He had kept a terse survey journal, a sort of log book, with columns for date, compass heading, miles travelled, and brief remarks:

Frydy   14   NNW   2   at the End of this Came
                                            to the Great Cave one of
                                            the Greatest Raritys the
                                            Country affords. [JS, 13]

The remarks tended toward abbreviation and employment of trite adjectives of only vague descriptive value, like "pretty," "pleasant." The remarks are terse: "this Dead water," "Some hard Rapids." He must have written the first manuscript with the survey journal at hand. It is written in bold script, corrected with interlinear additions and changes, and shows marginal additions that are incorporated in the second manuscript journal. He probably began the second manuscript (also in Carver's hand but in a much tighter and smaller script) at Michillimackinac and either finished it there or in Boston on his return in the autumn of 1768. On August 8, 1768, the *Boston Chronicle* announced his arrival by way of Fort Pitt. On September 12, it carried his advertisement for subscribers, at "two Spanish Dollars" a copy, to "An Exact Journal of his Travels," including accounts of the Indians, particularly the Sioux, of the interior lakes and products of the country, "Draughts and Plans" of the country, and "curious figures" of "Indian tents, arms and of the Buffaloe Snake which they worship." As early as September 24 of the previous year he had written his wife a letter containing several of the incidents later to appear in the "Journal." That letter was also published in the *Chronicle* (February 22, 1768), and confirms that even the long section on the Indians, the section in the *Travels* most larded with borrowings from other authors, was Carver's idea and not some printer's. In short, Carver had his book well in mind before he left Boston. When he sailed for England (February 22, 1769) aboard the *Paoli*, the *Essex Gazette* (Salem, Massachusetts) reported that he went with "Draughts, journals and good recommendations for his faithful service." He had every reason to expect that he was "on his way." That his book failed to appear for a decade is another story—one of frustration, poverty, official indecision and, of course, the intervention of that "unhappy division between Great Britain and America," as Carver called the Revolutionary War (T, vii).[14]

About his trials publishing his "Journals" once he was in England, Carver gives an account, corroborated in several particulars by documents and testimony of others:

On my arrival in England, I presented a petition to his Majesty in council, praying for a reimbursement of those sums I had expended in the service of government. This was referred to the Lords Commissioners of Trade and Plantations. Their Lordships from the tenor of it thought the intelligence I could give of so much importance to the nation that they ordered me to appear before the Board. This message I obeyed, and underwent a long examination; much I believe to the satisfaction of every Lord present. When it was finished, I requested to know what I should do with my papers; without hesitation the first Lord replied, That I might publish them whenever I pleased. In consequence of this permission, I disposed of them to a bookseller: but when they were nearly ready for the press, an order was issued from the council board, requiring me to deliver, without delay, into the Plantation Office, all my Charts and Journals, with every paper relative to the discoveries I had made. In order to obey this command, I was obliged to re-purchase them from the bookseller, at a very great expence, and deliver them up. This fresh disbursement I endeavoured to get annexed to the account I had already delivered in; but the request was denied me, notwithstanding I had only acted, in the disposal of my papers, conformably to the permission I had received from the Board of Trade. This loss, which amounted to a very considerable sum, I was obliged to bear, and to rest satisfied with an indemnification for my other expenses. [T, xii–xiv]

He had reentered complex culture and began the test of his "faith."

For more than a decade Carver struggled in England to interest the government and the public in his dreams for the Midwest and struggled, as well, to subsist. "In 1773 he petitioned to be made agent for Indian affairs in the region" he had explored; in 1774, according to his own remarks in the *Travels*, he nearly convinced "Richard Whitworth, Esq. member of parliment for Stafford" to search with Rogers and himself for the Northwest Passage along the route he had attempted; and in these years he attempted to interest " 'a private society of gentlemen,' possibly the Royal Society," in a journey across Asia to North America. All came to nothing, and on the eve of the publication of his *Travels* in 1778, he was apparently in such financial trouble that he appealed to Joseph Banks "to ask for some further assistance" to "save me from ruin," having, apparently, already been aided by him. The *Travels* were finally published, "but the income did not sustain the author, and in 1779 he was employed as a lottery clerk." The book was a success, and a second edition was printed in 1779 as were his pamphlet on tobacco culture and a geographical-travel anthology to which he lent the growing fame of his name. He oversaw the corrections and the beginning of the third edition, according to one source, but before it appeared he died of starvation at the age of seventy.[15]

The details of Carver's life in London are ironic—the stuff of dramatic biography. Encouraged by his culture to collect news of nature especially pertinent to the spread of empire, he was rewarded so insubstantially that he died a victim of the culture's economic system. Carver's private agonies reflected and flowed from unresolved tensions in the culture, a culture that was evolving a system to replace private patronage, and through which to channel support for projects it generated. But it could not yet adapt the new system to all contingencies. Military and Royal Society "patronage" had been organized to mount three voyages to the South Seas under Captain Cook; but Carver, a lone reporter, could not marshall similar support for his new project, or even to defray the cost of publishing a report of his last. Without personal fortune or adequate private or public patronage, he chose the only avenue left—the popular press—and singly exploited his knowledge. It is no wonder, then, and certainly no shame that Carver showed an equal interest in charming his reader as in informing him. Indeed one may argue that by accepting the challenge to operate doubly as a raconteur and reporter, he produced a book at once delightful and instructive. He resolved, *ad hoc,* in his narrative the tension between the demands of the two systems of reality he had internalized—those of the "cosmos of culture" and the new "cosmos of nature."

Coping with Nature and Culture

In his *Travels,* Carver sought to convey persuasively the good news of nature to a mixed audience of literati, virtuosi, lay readers and bureaucrats; English and American; adept and naive. That became his problem. There was as yet no satisfactory generic model to guide him, to help him make choices about diction, composition, tone and format—and no help from that stylized anticipation of audience response that supports an author choosing to work within the limits of a conventional genre. Had he been less ambitious and tried merely to report the facts and let them delight when they could, or to entertain and in the process bootleg some instruction, he could have drawn on one or another well-established genre for guidance: the Royal Society essay, the travel narrative, the catalogue, the promotion, the epistle, the journal. But Carver was ambitious and sought to convey all the news he could (both fresh and secondary) and at the same time inspire others with the fullness of his vision. His struggle to negotiate the mazeway generated by this meeting of his untamed intent and his readers' conditioning, displays in frozen facsimile the politics and drama of coping with culture in a time of changing patterns of fact and value, form and content.

Carver did submit without much struggle to certain conventions of the day. His audience expected and received the ordinary mix of accoutrements: maps, glossaries, illustrations, index, dedication, introductory apology: "And here it is necessary to bespeak the candour of the learned part of my Readers in the perusal of it, as it is the production of a person unused, from opposite avocations, to literary pursuits. He therefore begs they would not examine it with too critical an eye" (T, xvi). Not only the sentiment but even the diction and syntax are wholly conventional, not Carver's native voice. Most stilted of all is the flattering rhetoric of his dedication to Joseph Banks, Esq., President of the Royal Society:

Sir,

When the Public are informed that I have long had the Honour of your Acquaintance—that my Design in publishing the following Work has received your Sanction—that the Composition of it has stood the Test of your Judgment—and that it is by your Permission a Name so deservedly eminent in the Literary World is prefixed to it, I need not be apprehensive of its Success; as your Patronage will unquestionably give them Assurance of its Merit.

More usually, as within the body of the book, Carver draws on one or another rhetoric as the occasion seems to demand. Speaking of wild indigo, he adopts the plain style of the Royal Society: "*Wild Indigo* is an herb of the same species as that from whence indigo is made in the southern colonies" (T, 520). At other times he evokes the sublime, as when he writes of the waters of Lake Superior which he assumes drain beneath the Straits of Mackinac: "they must find a passage through some subterranean cavities, deep, unfathomable, and never to be explored" (T, 142). One is reminded of how Samuel Taylor Coleridge, who knew Carver's work (CHAL 1: 194), developed a similar image in "Kubla Khan": ". . . Alph, the sacred river, ran / Through caverns measureless to man / Down to a sunless sea." Carver is most himself and most convincing when he "forgets himself" and writes vivid and particular capsules of reportage. He says of the puff adder, "when any thing approaches, it flattens itself in a moment, and its spots, which are various dyes, become visibly brighter through rage" (T, 167–68). Similar rhetorical tactics also characterize the prose of John Bartram, Mark Catesby and other contributors to the literature.

Of course Carver borrowed the journal form for the first part of the *Travels* ("A Journal of the Travels, with a Description of the Country, Lakes, &c.," 163 pages), the informal essay for the second quite separate and long section ("Origins, Manners, Customs, Religion, and Language of the Indians," 261 pages), and the encyclopedic catalogue for the concluding major portion which describes specific birds, reptiles, fishes, trees, herbs and flowers and arranges them in alphabetic order within categories (85 pages). An author must start somewhere, but within each section Carver copes with the demands of his task and apportions to one section or another, as it suits his purpose and sense of belonging, the intelligence he means to transmit. Like the rhetoric, the content of Carver's *Travels* is also mixed. Much of the geography and natural philosophy is original; that is, Carver is reporting what he actually saw or experienced, as, for example, his view of St. Anthony Falls on the Mississippi, his drawing of which is the first ever published. Some material, like his description of the beaver, is a mixture of his own observation and his interpretation of what had been said and thought by others, such as Robert Rogers. A sizable portion of the book is hearsay, and he so indicates, crediting an Indian, a trader, a common legend or another author. He does not claim, for instance, to have seen the headwaters of the "River of the West" (T, 76) but frames his reports with words like "these Indians informed me" (T, 117) and "from the intelligence I gained" (T, 76). These tales add a dimension to the book that would have been lacking had Carver reported only what he had personally experienced or discovered. They seem not intended to deceive, but to gather in all the relevant evidence. They form an integral part of the whole and are not mere padding or plagiarism as some have claimed. There are portions of the book, especially in the Indian section, that are rather completely "borrowed" from other authors. The authors are alluded to, but specific incidents and phrases are not always framed by quotation marks to indicate clearly what is Carver's and what is taken from Hennepin, Charlevoix, Lahontan and Adair. But it is relevant to point out that all nature reporters mixed personal and published materials. One could find in Jefferson's *Notes on Virginia* similar instances of tacit quotation; certainly in the works of Sir William Johnson, Robert Rogers, Cadwallader Colden and Mark Catesby; and they abound even in the nineteenth-century works of Lewis and Clark, Audubon and William H. Keating.[16]

Carver meant to deliver to the reader as full an account as he could and worried little—sometimes too little—about the smaller details of authorship. With the manuscripts of Carver's "Journals" in the British Museum is a note "To the Reviser" in Carver's own hand, which, in addition to clarifying certain terms and offering to correct "ambiguous" phrases and supply additional information, appears to give *carte blanche* to the reviser: "any thing that the Reviser shall see fit to add to Embellish or give better sense to the Journal will I dare say be very agreable to the Publishers and to the Author" (J2, 53). This willingness to submit to massive editing is typical. Many explorers went even further, relinquishing entirely to others the task of composing the narratives of their explorations, as William Clark did to Benjamin Smith Barton and Nicholas Biddle, as Stephen H. Long did to William H. Keating. Such accommodating pliability does not, however, mean that all editing was done by another hand. Carver spent no little energy himself in correcting, rephrasing, rearranging, adding, emending and polishing his manuscripts through their several versions.

Some of Carver's particular and characteristic compositional choices are illuminated by comparing his manuscripts with the *Travels,* something that could not have been done by Bourne and earlier critics who lacked the manuscripts. The material covered by the manuscripts is the narrative of his travels, but within the manuscript narrative are items that Carver subsequently relocated to the middle section on Indians or the final section on flora and fauna. The first manuscript journal (J1) runs the equivalent of forty-nine pages of double-spaced typescript. The second (J2) runs to forty-six, but to it have also been added thirty-four pages of notes, anecdotes, drafts of introduction and notes for insertion in places identified by marginal numbers and emphasized by small drawings of hands with pointing forefingers. What is interesting is that Carver had begun, even between these two rather rough drafts, to edit and tighten the narrative. In the first version he took two full pages to detail his travels west from Boston to Michillimackinac. He cut the second version to a page and a half, but it was still in rough form and filled with detail not essential to the story he had to relate. The first two sentences will suffice to indicate the texture:

May 20th 1766   Set out from Boston in New England, with design to make discoveries to the Northward and Westward, and if practicable find a northwest passage. 29th reached Albany and June 13th Oswego from whence in company with some traders bound to Detroit, came the 23d to Niagara, 26th to Fort Erie and July 13th to Detroit. [J2, 1]

Here surely are what Samuel Johnson called "such enumerations as few can read with either profit or delight." In the *Travels,* finally, the Boston to Michillimackinac journey is reduced to one paragraph:

In June 1766, I set out from Boston, and proceeded by way of Albany and Niagara, to Michillimackinac; a Fort situated between the Lakes Huron and Michigan, and distant from Boston 1300 miles. This being the uttermost of our factories towards the north-west, I considered it as the most convenient place from whence I could begin my intended progress, and enter at once into the Regions I desired to explore. [T. 17]

Did an editor's blue pencil accomplish this last condensation? No one knows, but I see no reason to infer it did. The two manuscripts date from 1767–69, leaving Carver nearly a decade to continue the cutting and tightening he had begun almost at once. By the time he went to press Carver had spent years telling his story to friends and to the Plantations Office, had copied his manuscripts several times and had read other travels. He could condense by himself since he had freed himself from the tyranny of details through a decade of practice at "getting on with it."

A great deal of material appears in the *Travels* for which there is no manuscript copy. Someone added it; again, who is not known. But adding was one of Carver's characteristic devices, as the marginal additions to the first manuscript indicate and as the thirty-four pages at the end of the second further demonstrate. Here one finds recognizable portions of anecdotes that appear in the *Travels*; for example, "the Indian and his rattle snake," some of the Introduction, rattlesnake-bite treatment and some of "An Address to the Public" (J2, 68–69, 54, 70, 69). That it was also his intention to add material from standard sources and incorporate it into his *Travels* is shown by comparing his treatment, in the second manuscript and in the *Travels*, of an "extraordinary" phenomenon at Detroit. Carver had passed through Detroit in 1766, but his *Travels* includes the report that, "in the year 1762, in the month of July, it rained on this town and the parts adjacent, a sulphureous water the colour and consistence of ink . . ." (T, 153). In the manuscript a blank appears where the date should be: "At Detroit in   the day become as dark as night at the same time it rain^d water as black as ordinary ink which continued for some hours" (J2, 65). Apparently Carver had been aware while writing his manuscript that he would be able to find a published account of this occurrence and insert the data at a later time; the "remarkable Darkness" had been reported in 1763 as "An Account of a remarkable Darkness at Detroit in America . . ." (RSPT 53: 63–64). He did not add this detail with an intent to deceive; he did not claim to have been in Detroit in 1762. He did it, one must conclude, because he was adding material that would expand the book's usefulness, just as he was to do in so many other instances which were later viewed as plagiarism. Similarly, in many instances he filled out his earlier rough drafts to make a more complete report. In the second manuscript one reads of a Sauk village: "Their produce, great quantities of indian corn, beans, squashes, melons, indian tobacco &c." (J2, 13); in the *Travels*, "a great quantity of Indian corn, beans, *pumpkins*, squash, and *water melons*, [italics added] with some tobacco" (T, 37). Clearly, in rewriting he retained the original list but recalled additional details and inserted them.

Even more significantly, Carver tinkered with his expression—diction, syntax, imagery. He often edited with an eye and an ear to conventional standards of expression, displaying the ways in which his own complex of literary values gave obeisance to notions of decorum not formed on the frontier. Some of the changes may have been unfortunate and have diluted the vividness of his less self-conscious earlier expression, but the changes were his and not those of an editor. He says of the July weather at Grand Portage (Lake Superior) in the first manuscript, "the weather so Cold I Could ware a Coat and a Jacket and a Cloak in the Middle of the Day" (J1, 79); in the second manuscript, "the weather so cold I could wear a cloak comfortably over winter cloaths in the middle of the day" (J2, 45). By dropping the series, adding the adverb "comfortably" and inserting the general phrase "winter cloaths," he made the second version conform more closely to standards of eighteenth-century expression, but he also weakened the force of the original report's implication of surprise and discomfort at the cold. The first suggests that he *must* wear "a Coat and Jacket and a Cloak;" the second, that he *might*.

There are, as has already been noted, some remarkable descriptions in Carver's *Travels*. At their best they forcefully involve the reader through Carver's personal participation in the scene, as when he describes the coldness of Lake Superior water: "the water drawn from thence was so excessively cold, that it had the same effect when received into the mouth as ice" (T, 133). His best descriptions thus call upon the senses for their imagery. That such sensuous details are not the additions or suggestions of an editor, but reflect Carver's sensibilities, is demonstrated over and over in the manuscripts. Indeed, what one finally becomes aware of is that the manuscript descriptions are even more vivid and concrete than the *Travels'* descriptions of the same things. Apparently an inner censor operated to smooth and generalize and to recast into visual and auditory images, with their necessarily more remote perspective, the more concretely felt images of his first invention. The most remarkable example of this self-editing process can be seen by comparing his three accounts of the Great Cave. The published version is the best known and serves as an adequate introduction to the cave:

The entrance into it is about ten feet wide, the height of it five feet. The arch within is near fifteen feet high and about thirty feet broad. The bottom of it consists of fine clear sand. About twenty feet from the entrance begins a lake, the water of which is transparent, and extends to an unsearchable distance; for the darkness of the cave prevents all attempts to acquire a knowledge of it. I threw a small pebble towards the interior parts of it with my utmost strength: I could hear that it fell into the water, and notwithstanding it was of so small a size, it caused an astonishing and horrible noise that reverberated through all those gloomy regions. I found in this cave many Indian hieroglyphicks, which appeared very ancient, for time had nearly covered them with moss, so that it was with difficulty I could trace them. They were cut in a rude manner upon the inside of the walls, which were composed of a stone so extremely soft that it might be easily penetrated with a knife: a stone every where to be found near the Mississippi. The cave is only accessible by ascending a narrow, steep passage that lies near the brink of the river. [T, 63–65]

In all three versions of this scene, the cave is approached from the river, entered and explored by Carver who personally throws stones to test its depth, records the reverberations of the sound, and notes the ancient aspect of the moss-covered hieroglyphics carved into the soft sandstone. The reader is taken into the scene by identifying with the persona, Carver. But surprisingly, while the published version is as mysterious and evocative as any romantic could wish, and includes phrases like "unsearchable distances," "gloomy regions" and "rude manner" that remind one of the growing romanticism of the late-eighteenth-century poets and painters, the earlier two versions are even more evocative. Carver has not applied sensational and picturesque details to an otherwise plain report, as might have been guessed from the *Travels* alone, but just the reverse; he has sacrificed "felt knowledge" to what he takes to be literary felicities of expression and exposition.

In the first draft, Carver "Cast a Stone which I Could hear fall at a distance and with a strange hallow sound"; he "tasted of this water and found it to be very good"; he records "Frightfull appearances of Lights shining at a distance and strange sounds" when the cave is explored by canoe; and he describes the rock of the bluff in which the cave exists as "very soft Like the grit of a Grind Stone" (J1, 49–50). These images—appeals to the reader's senses of hearing, taste, sight and touch—make the description concrete. To these he adds the eighteenth-century "sense" of intellectual awe of antiquity by reporting his addition of "the Arms of the King of England" to the "Strange Hieroglyphicks . . . Very ancient and grown over with moss."

In his second rendering of this scene, Carver retains the "hollow sound" and "lights shining" but omits the tasting of the water, and oddly destroys the sense of touch in his image of the sandstone by calling it "soft as grindstone," perhaps through miscopying (J2, 18–19). The hieroglyphics remain. He has, however, expanded a remark about the Indians from the first version into a generalization on their superstitious nature: "The Indian say they have attempted to go with a canoe and light upon the water but have been deterred by frightful appearances of lights shining at distance and strange sounds which they think produced by spirits as every thing not easily accounted for by them always is" (J2, 19). This generalization has the effect of directing the reader's attention away from the cave and its mystery to Carver, the reporter on Indians. Its tone shows him as aloof, detached and speculative rather than involved. In the published version of the *Travels,* this unfortunate intrusion has been removed and appropriately deposited in a chapter on Indians, "Of *their* Religion" (T, 380–89). Even so, excellent as the published version is as an expression of Carver's intimate, personal, sensuous, spiritual and intellectual participation in nature, it standardizes, generalizes and omits the taste report: "clean white sand" becomes "clear sand"; the lake which in the manuscript is so "dark I Could not find out the Bigness nor form" becomes in the *Travels* "a lake, the water of which is transparent, and extends to an unsearchable distance" and the sandstone, first "very soft Like the grit of a Grind Stone," certainly a vivid, tactile simile, becomes "a stone so extremely soft that it might be easily penetrated with a knife: a stone every where to be found near the Mississippi." This last image is still concrete, but less immediate, subjunctive in mood and capped by a generalization that softens its particularity.

What this analysis of the several versions suggests is that— even though Carver explored the interior of the country and recorded novel phenomena with an attention to detail and facility of language refreshing in the eighteenth century—he failed sufficiently to grasp the stylistic implications of his location at the nexus of competing rhetorics, failed to realize the full dimensions of his opportunity to legitimate a new voice in American literature. That was not of course his mission as he understood it. He tinkered with his style and compromised tactically so as to reach a wider audience, a result he demonstrably achieved. What effect Carver's style had on sales is problematic: maybe none, probably some. Still Carver shared with other nature reporters and probably with most of his readers a lack of interest in style as such. Nature itself and not the medium interested them; style was artificial, something one applied for decorative effects or to elevate the tone when eloquence seemed in order. Nature reporters often lacked the self-assurance and monomania of self-conscious artists who strive to bring even the minutest details of their expression into harmony with the implications of their form and content.

Carver's most significant successes are strategic rather than tactical for he discovers a formal device in his own inclination to add details, digress, embellish. It is an ethical solution with esthetic consequences, allowing him to retain the armature of factual geographical-chronological progress and yet add whatever anecdotes, philosophical asides, promotional puffs or useful "applications" he can think of or discover. He does not fabricate a persona which then controls the talk from within the *Travels,* as novelists and poets increasingly came to do. He controls from without, exploiting nature or his own experiences as data to manipulate.

This ethical pivot makes nature reportage a rich source of cultural intelligence, allowing the reader to infer a good deal about the author's tacit knowledge of the ways that fact and fiction seem to cohere (or fail to cohere) in the culture the author shares with his audience. The fictions of a culture are what lend value and meaning to facts, turn a stone into ore, a comet into a sign, environment into "place," movement into "pilgrimage," Indians into "children of the forest." What one discovers by looking closely at both the smaller and larger rhythms of the *Travels* is an artificial and complex web of interconnections between the "cosmos of nature" and the "cosmos of culture." Carver more than he could know became an agent of the culture and helped objectivate the still sometimes tenuous connections between the new nature and civilization. An emerging cosmos requires numerous and particular new micro-connections to be made as well as new links between them and older patterns of long service—religious, civil, domestic, affectional. Poets and philosophers perform this function at a high level of symbolic abstraction; vernacular reporters work hip-deep in the "lumber" of the world.

At the simplest level Carver links fact to fact, and fact to significance by hanging from the chain of the daily log entries whatever clusters of associations one link or another suggests. In the second manuscript Carver records briefly the remarkable plants, animals and Indians found at each stop; at Green Bay,

Fox River Portage, Prairie du Chien. At Green Bay there is "a kind of shrub about the pitch of sweet fern bearing a very delicious fruit about the bigness of a small musquet ball and very black when fully ripe, weighing down the bushes into the sand and therefore called sand cherries. There are other wild fruits in these parts, as whirtleberries, blackberries, gooseberries, and currants" (J2, 5). In his *Travels* this item is developed into a short paragraph:

Near the borders of the Lake grow a great number of sand cherries, which are not less remarkable for their manner of growth, than for their exquisite flavour. They grow upon a small shrub not more than four feet high, the boughs of which are so loaded that they lie in clusters on the sand. As they grow only on the sand, the warmth of which probably contributes to bring them to such perfection, they are called by the French, cherries de sable, or sand cherries. The size of them does not exceed that of a small musket ball, but they are reckoned superior to any other sort for the purpose of steeping in spirits. There also grow around the Lake gooseberries, black currants, and an abundance of juniper, bearing great quantities of berries of the finest sort. [T, 30]

One can see exactly where Carver has expanded, rearranged, and added material, especially the size of the shrub, the French name of the cherries, a speculation about their growth, and a comment on their potential usefulness when steeped in liquor. Unrecorded in the manuscripts but following the account of sand cherries in the *Travels* is a two-page description of "sumack," which the Indians mix with tobacco and "which causes it to smoke pleasantly," and of "Segockimac," a vine which the Indians also mix with their tabocco (T, 30–31). Clearly, one thought led to another here (thoughts of drinking to thoughts of smoking) and reminded him of details not contained in his first manuscripts. He must have searched especially for items whose utility was transparent, as would any nature reporter alert to the message broadcast by the publications of the Royal Society and presses in America and abroad. This was simple but basic work. Reporters needed to do no more unless they aspired to grander effects, as Carver did.

The liberating effect of Carver's strategy appears more clearly on a larger canvas. In the journal portion of the *Travels*, Carver finds niches for large and small additions: whole five- to six-page digressions upon Indians or empire; numerous little holes to be plugged by reports of native copper in one place (T,138), gold in another (T, 135), giant trout in a third (T, 140). It is difficult in a limited space to convey persuasively the ultimately rhythmic effect of his practice, but it might help to outline one thirteen-page section containing anecdotes as well as particular reportage (T, 30–43).

This section opens with his two-page description of sand cherries and sumac around Green Bay, then spins out into a six-page account of the Winnebagoes. From the middle of this short ethnography Carver loops a tale told him by an elderly chief, a kind of epicycle on an epicycle, each one more enchanting than the last. In this he relays the chief's tale of an encounter "forty-six winters ago" with a Spanish caravan "several moons" to the south, the clincher of which is that the Indians relieved the Spaniards of a "ponderous burthen" of silver but not knowing its worth, abandoned it in the woods. Then leaving the Winnebagoes, Carver describes the topography, flora and fauna of the Fox River region, but digresses again with his tale of the Indian heroine's escape from captivity. Then he goes back to a description of the lakes and portages between the Fox and the "Ouisconsin" rivers. Thus Carver varies the content and demonstrates his ability as a storyteller as well as a reporter of factual and specific details of geography and biology. Reading the "Journal" portion of the *Travels* straight through, one becomes aware of a kind of rhythmic contraction and expansion, contraction to mere factuality for several pages, followed by expansion into speculation, digression or anecdotes of relevance and delight, then contraction to solid reportage again. Neither the facts nor the fictions are subordinate. The chief drama here is the struggle between fact and fiction, a drama that binds Carver to later artists like Whitman and Melville who similarly heaped up facts to lend substance to their fictions, then invented fictions to control and give meaning to the catalogue of facts which was the world, was America. This comparison, however, must not be pushed too far for they were artists with a different and more self-conscious command of strategy. The thrust of these remarks is not to demonstrate any "influence" of the nature reporters on later literary artists, a tiny influence when compared to the effect of giants like Carlyle, Wordsworth and others, on Whitman and Melville. The point is instead that like them, Carver coped with a problem of American culture that would occupy the attention of the best authors America produced in the nineteenth century. What did nature have to say and give to man—commodity only? Or a design for a new ethic? How should one live in the new arena of North America? How should one civilize nature or naturalize culture when whole systems of social and intellectual tradition had become explicitly problematic with the rise of new civil and industrial institutions? Carver showed instead of told.

Carver's one traveller (moving along, through nature, past indigenous populations, toward a dream of inland and future civilization) personalized the conditions and drama of what was to become increasingly the American national/cosmic myth. James Fenimore Cooper adopted the theme and worked it in a succession of novels in which the type of Jonathan Carver re-appears in many guises; but that is another story. The concern here is Carver's compositional achievements in coping with the tensions between culture and nature. His device was certainly simple, having the appearance of nature rather than artifice as each detail appears to grow spontaneously, as in the oral tradition, out of the location being described and the associations of the speaker's mind. Of course it was not spontaneous, as the manuscripts reveal. Carver shifted whole packages of materials around until he found for them what seemed like an appropriate resting place; for example, he first intended to group his observations and speculations about the Indians around the winter of 1766–67 and his stay with the Sioux on the Minnesota River, but later shifted many of these materials to the Grand Portage section of the "Journal" or to Part II, the section on Indian customs. The form is artfully natural, not automatically so.[17]

This illusion of personal spontaneity and naturalness is Carver's best effect. But it is a fiction that sometimes hides un-resolved tensions, as for example the tension between Indian and non-Indian cultures. With this Carver could not cope; or rather, he coped too easily, resolving dissonance *ad hoc* when instead a high order of systematic synthesis was needed. Carver could not successfully integrate Indians as he perceived them into the cosmos. He could not decide whether Indians belonged to a "cosmos of culture" or the "cosmos of nature." Within the narrative portion of his *Travels,* for instance, Indians function largely as decoration, as animated landscape—just as gypsies and highwaymen do in conventional eighteenth-century narratives of the Grand Tour or of domestic travel through England and Scotland. Or alternately he placed them apart in a "ghetto" (the "reservation" of Part Two), exotic and foreign, like Tahitians. The Indians' ethnological variety could not satisfactorily be integrated into the eighteenth-century world view which took the variety of species in plants and animals as evidence that nature was designed to serve culture by filling the larder, inspiring awe, teaching lessons. Indian variety sometimes did this, but too often in the experience of colonial Americans, Indians declined to be accommodated by the religious, schooling, economic and military systems central to American culture. The treachery of "praying Indians," their "fickleness" as military allies, their manifest and notorious disinclination to become agriculturalists or to send their children to schools instituted for their "improvement" had long been self-evident to colonists. It had driven many to reconsider whether the Indians were wholly human. They seemed so at times. At other times they seemed completely alien, devils or animals. Carver could not heal this dissonance with the resources available to him. Neither could anyone else, not missionaries, not soldiers, not even natural philosophers of the Royal Society or poets of London or the colonies. Systematic integration into a cosmos containing both nature and culture had to wait at least a century: await geological reassessments of the earth's age; await the narrative devised by theorists of evolution to account for the development of animal and plant species. Only then, with the advent of modern anthropology, could cultural diversity be assimilated into the larger story of nature. But it was Carver and others like him who collected and reported the diversity that would challenge scholars to make the later synthesis. Forcing the issue would not have advanced anthropology. It would only have falsified the reporters' experience, and the one thing they had to contribute to the evolution of a new cultural constellation was their novel and suggestive experience in nature.[18]

The Americanness of Jonathan Carver

Obviously Carver's *Travels* is a very American book, and not only because the geography, the flora and fauna, the Indians and weather were American. Carver's strategies and tactics are American in a deeper sense. So is his humor, his enthusiasm, his characteristic rhythm and style. Culture is not simply environmental any more than nature is. It permeates the very workings of mind to mold what one comes to take as self-evident, satisfying, real. But nature is global and culture, regional so that while water may be H_2O everywhere, it is perceived differently by Yankees and Hopis. Similarly if less radically, when eighteenth-century Englishmen and Americans reported nature, they produced accounts of a discernibly national flavor. To illustrate this, we must turn briefly to some nature reportage similar to Carver's but not written by an American.

Anglo-American culture bred many like Carver who travelled over the land and seas to collect and return to London the facts of botany, ethnography, zoology, geography and navigation that were then published for the delight and instruction of an eager public. The greatest among these late-eighteenth-century travelling nature reporters was remarkably like Carver in several respects, though his fame and accomplishments far exceed Carver's. James Cook (1728–79), like Carver, found release from village origins and limited horizons through service to his king. He, too, volunteered as an enlisted man for service against France (1755) and rose to officer rank, though unlike Carver, Cook was kept in the service. Carver taught himself surveying; Cook was self-taught in the elements of mathematics and astronomical navigation. His excellent survey of the channel of the St. Lawrence River (1759) brought him to the attention of the Royal Society of London, which cosponsored his later voyages to the South Pacific.

During the decade of Carver's frustrations in London, Cook made the voyages that brought him contemporary fame and earned him his lasting reputation. At the request of the Royal Society, he commanded the *Endeavour* to the South Seas (1768–71) to view the transit of Venus. He was accompanied by Joseph Banks (later President of the Royal Society), the man to whom Carver was to dedicate his *Travels,* and by others well equipped to report on nature: Dr. Daniel Solander, botanist; Mr. Alexander Buchan, artist; Sydney Parkinson, painter. He made a second voyage (1772–75) to discover or disprove the existence of a great southern continent in the South Pacific. It was from

this voyage that Captain Furneaux of the *Adventure* returned
with Omai, a Tahitian who immediately became the fashion
of the London literati as the embodiment of the "noble savage."
In 1776 Cook began his third voyage in a decade to investigate
specifically that most captivating of English dreams, a northwest
(or northeast) passage to the Orient. He returned Omai to
Tahiti, scotched the existence of the northwest passage and died
the victim of the natives of the Sandwich Islands. One commen-
tator, Sinclair H. Hitchings, says of Cook's voyages, "Captain
Cook's three voyages to the Pacific are among the great scientific
adventures of all time," and notes the "intense popular interest"
in the "scientific achievements and romantic and heroic epi-
sodes" of the voyages. Reading the magnificent quarto and folio
volumes of Hawkesworth, Parkinson and Cook that recount the
voyages, and overflow with maps and engravings of islanders,
plants, fish, landscapes and all varieties of natural history, one
can understand the enthusiasm and delight that these voyages
stirred. They combined the sweep of Hakluyt with the detail
and accuracy of the Royal Society of London *Philosophical
Transactions*.[19]

Clearly Carver and Cook served the "nature project" similarly. As reporters they mapped and recounted the particulars
of nature. As authors they delighted the virtuosi and public with
pictures of exotic peoples and faraway places, even though their
chief aim had been to convey objective facts in plain prose.
How then is Carver's *Travels* American? Is it merely a colonial
specimen of an Anglo-American type, simply a colonial Englishman's response to the interior of North America, or is it an
expression that reveals distinctly American qualities? And which
ingredients are common to Anglo-American nature reportage?
Which are peculiarly English or American?

To one who would celebrate Carver's importance as an author of travels, the initial exposure to the accounts of Cook's
voyages is sobering. Cook's voyages were of far greater consequence, scientifically, geographically, ethnologically and mili-
tarily than were Carver's. The reportage is generally more re-
liably accurate in detail and comprehensive in scope, especially
the mapping. The illustrations of specimens and the engravings
of landscapes and natives are vastly superior to the relatively
crude illustrations in Carver's work. Even the format, binding
and sheer bulk of the multi-volume voyages overshadow Car-

ver's poor octavo volume. And Sir Joseph Banks's *The Endeavour Journal 1768–71* is so full, so felicitously expressed and so packed with learning that Carver's descriptions of birds and plants seem amateurish by comparison. Thus, exposure to the narratives of Cook's voyages impels me to sympathize with historians who dismiss Carver as a minor figure of little consequence.

Nevertheless, looking more critically at the Carver and Cook narratives, I find significant differences, not all to Carver's disadvantage. The English accounts of Cook's voyages suffer as literature from their very excellence as full journals. It is difficult to identify with any one person in Cook's accounts. The usual pronoun is either the third person "he" or the first person, plural "we" instead of Carver's "I." Also, the English accounts maintain so faithfully a linear journal form that long passages are filled with exactly the minutiae of tides, winds, latitudes and locations that "few can read with either profit or delight." The tone is generally impersonal and aloof, so that while I admire the accuracy and objectivity of descriptions, I remain relatively unmoved, or am moved only by the object described rather than seduced into the scene by the author's involvement.

This difference in tone is crucial. Carver invites the reader to participate in nature, to share his Crusoe-like adaptability to moose-lip cuisine and sumac-laced tobacco. The landscape and climate are not extravagantly alien, the flora and fauna are not so strange as to stun the imagination. With Cook, on the other hand, we meet exotic and thoroughly foreign climes and bizarre flora and fauna as in a travelogue. An adventuresome American could set off in a canoe and follow Carver but could not so follow Cook. A reader may identify with Carver, as one does with Natty Bumppo, and become vicariously an insider with Indians, trappers and explorers, and revel in living off the land. With Cook the reader remains an outsider, conscious of European superiority and responsibility, amused and piqued by Tahitian thievery, and part of a large project. That this difference between the tone of the two is national, not just circumstantial, is bolstered by an account of Cook's third voyage by that irrepressible, romantic and imaginative Yankee, John Ledyard.

John Ledyard (1751–89) was born in Groton, Connecticut,

and through a variety of circumstances came to ship with Cook in 1776 on the third voyage as a corporal of marines. On his return to America (1782), Ledyard deserted the British Army to avoid fighting fellow Americans, and rushed his narrative of Cook's third voyage into print before the official accounts were published. He borrowed occasionally from an even earlier unauthorized account but wrote from his own journals when reporting events in which he had participated. Apparently nearly everybody aboard had kept journals with an eye to publication, knowing the eager market that awaited such accounts. Like Carver, Ledyard was eager for personal adventure and was driven by such goals as fur trade with China and walking across Russia. While his book was being printed he travelled up and down the East Coast to arouse interest in and obtain backing for the northwest fur trade. Failing, he went to Europe where, according to Jefferson, he intended to walk across Russia to the "Northwest coast of America, and to penetrate through the main continent to our side of it." Jefferson remarked that Ledyard intended to be the first "circum-ambulator of the earth" and that "having no money they kick him from place to place and thus he expects to be kicked around the globe." He made it as far as Irkutsk before he was arrested and returned to Europe.[20]

Ledyard's narrative expresses his enthusiasm for adventure, and his account is personal like Carver's. He adopts the first person singular pronoun to recount his own experiences. He says of a journey in a kayak, "there was no other place for me but to be thrust into the space between the holes extended at length upon my back and wholly excluded from seeing the way I went or the power of extricating myself upon any emergency" (pp. 93–94). In another place he reports an Alaskan breakfast he endured: "it was mostly of whale, sea-horse and bear, which, though smoaked, dried and boiled, produced a composition of smells very offensive at nine or ten in the morning" (pp. 96–97). In several places he is critical of Cook's lack of understanding and sympathy with the natives, reminding one of Carver in his more romantic moods. Finally, however, unlike Carver who was forty-one years his senior, Ledyard is thoroughly romantic in sentiment and expression when describing natives and landscape, reminding the reader more of William Bartram than of Carver. In one instance, reporting the desertion of a sailor who ran away with a New Zealand native girl and was later captured by Cook, he says, Cook "surprized them in a profound sleep locked in each others arms, dreaming no doubt of love, of kingdoms, and of diadems," and concludes with a remonstrance to the god of love, "God of love and romance! this pair ought to have been better heeded by thee . . ." (p. 20).

Ledyard's *Journal* suggests, if it does not conclusively prove, that by Cook's time a split had developed between English and American treatments of the nature reporter's travel narrative. His example underlines, along with Carver's, the tendency of American travels to report adventures of men chafed by the "Establishment," proud of their self-sufficiency and attentive to the details of landscape on their passage to El Dorado, Paradise or Arcadia. His and Carver's examples underline, by contrast, the responsible, sober fullness of Cook's reports of his thoroughly financed, carefully planned voyages. The headlong rhythms, loose form and personal tone of the American narratives suggest their debt to an antecedent, native oral tradition; the style and tone of the English voyages suggests instead a heritage of official, written reports to the "home office."

I am moved, finally, to infer a divergence of American and English esthetics, at least in the realm of travel narratives, that extends beyond style into the very anatomy of imagination. Carver is engaged by sand cherries and by the Falls of St. Anthony, on one hand, and excited by potentials for asylum in the distant West or passage to the Pacific, on the other. But he is vague about the intervening steps to connect the two, feeling apparently more excitement in yoking East and West, the ordinary and the sublime, than in actually completing the tedious steps and accurate maps necessary to achieve such a union. His is a metaphysical joy at juxtaposing the earthly and paradisical, an inheritance of Puritan sensibility perhaps. In the English voyages I sense instead a controlled and balanced appreciation of the mundane and of all the intervening steps on the ladder to truth, and especially an esthetic appreciation of completeness and fullness. Finally, then, I take Carver more as the antecedent of Mark Twain than of Lewis and Clark, William H. Keating or Henry Schoolcraft, and see Cook as the parent of subsequent geographical expeditions of large scale: Lewis and Clark, the voyage of the *Beagle,* antarctic explorations and the conquest of Everest. Carver belongs to eighteenth-century American culture; Cook, to modern international science.

IV

John Bartram,

A Pennsylvania Farmer

Bartram Perdu

John Bartram is a problem, a mixture of transparency and obscurity. Unlike Franklin or Carver or Dudley he is hard to catch, partly because of what he did not write and partly because of what he did. He did not much care for writing as a way to make his mark. He preferred to work directly in the field and devote himself to nature reportage of a wholly objective sort, the actual collecting of seeds and specimens which he then dispatched to Europe and other colonies. Like many field biologists since his time, making systematic and sustained verbal representations of what he learned and had to teach took a back seat to what he did best, was paid and appreciated for doing, and was happiest doing day by day; namely, tracking down, noting carefully, and collecting specimens of nature. Yet he did keep careful logs and journals and maintained an extensive trans-Atlantic correspondence, and along the way wrote some remarkable pieces. These give us a glimpse of the vitality and entirely human being of the man, and allow us to reconstruct some of what he felt and how he negotiated the overlapping mazeways of culture and nature at a level deeper than he cared as a rule to expose.

The whole Bartram has been especially difficult to grasp. Rather, fragments have been saved and allowed to stand. Historians of science have picked up one piece, historians of literature another, and biographers a third. Often he has been presented as half of a father and son pair, John and William. Or else Bartram as a type has been puffed into a myth, as by Crèvecoeur. And to a degree, Bartram has fallen victim to disciplinary specialization and its taken-for-granted expectations. His prose disappoints literary scholars; his everyday being is largely beside the point for historians of science who care more for what he contributed to its advancement than for the particular ways he made sense, word by word, or laid down his feelings behind and beneath the words. And yet, approached alertly and from the advantage of one looking for nature reportage rather than art, Bartram comes slowly into focus as a reporter who has much to teach us about the difficulties and delights of being an observant participant in the life, both cultural and natural, of colonial America.

Of all the men who "numbered the streaks of the tulip" in America during the eighteenth century, few have been more admired by their contemporaries and praised by posterity for their contributions to natural philosophy than John Bartram of Pennsylvania. Linnaeus called him "the greatest natural botanist in the world," and travelling naturalists like Peter Kalm sought out Bartram when they came to America. Colonial virtuosi like Dr. Alexander Garden of Charleston and Cadwallader Colden of New York corresponded with Bartram regularly, invited him to visit them on his travels, and exchanged information, seeds and specimens with him whenever possible. He funneled some excellent epistolary descriptions of nature through Peter Collinson to the *Philosophical Transactions* of the Royal Society of London, but was most appreciated for his service to natural philosophy as a collector and distributor of novel and typical specimens to virtuosi in America and abroad. In recognition of his work, the Swedes honored him with membership in the Royal Academy of Science, and George III of England appointed him "Botanist to the King." [1]

No virtuoso other than Franklin has been more frequently memorialized in print than Bartram. At the same time, his writings are rarely anthologized, as are those of his son William, his friend Franklin or even his travelling companion Conrad Weiser. He moved no audience the way Carver did. The causes of this odd mixture of fame and neglect—fame for the reporter, neglect of his reportage—are several and complex, but explicable in terms of dynamics of culture and personality.

Bartram chose to work diligently within the preparadigmatic but emerging discipline of biology and to address that special audience rather than the larger public Carver courted.

The developing nature establishment recruited, sustained, exploited and rewarded Bartram's field work, and he internalized what he took to be their values and beliefs, their avowed regard for facts, their casual neglect of style, their systematic disregard of personality. Bartram elected to settle in and make himself useful to the system, bit by bit, rather than cut loose like Carver. He remained in the colonies, farming and botanizing, and trained himself to become the type of nature reporter most valued by the system he elected to serve. Many of his strengths, most of his flaws, and most of the treatment accorded him by his biographers are functions of this relationship to the system.[2]

For Bartram to "make it" as he did, up from modest beginnings and against economic and social handicaps, was a remarkable accomplishment, and one the memorialists and biographers have not tired of celebrating. But they have sometimes worked too hard to explain away his idiosyncrasies. John was born to Quaker parents (William and Elizabeth Hunt Bartram), March 23, 1699, by the Quaker calendar. When his father married his second wife and moved to North Carolina complaining of "wrongs and abuses" suffered in a "row" with the Darby Meeting, John was left to the care of his grandmother in Pennsylvania. His father was killed by the Indians in North Carolina. Some biographers have fastened on the killing and the row to explain John's later dislike of Indians and resistance to Quaker discipline. This is reaching. But worse, it is a kind of insult to Bartram, a refusal to take seriously his mature ideas and values, regarding them as flaws to be excused or extenuated instead of scrutinized in the context of his adult experiences, beliefs and view of nature. Deistic apostasy was no rare occurrence among natural philosophers in the eighteenth century and was a likely if not inevitable consequence of assenting to the view of nature Bartram embraced. About the Indians Bartram may indeed have harbored some unpleasant associations, but his most vehement anti-Indian expressions date from the period of the French and Indian Wars when the Indians were an active threat to the security of eastern Pennsylvania farmers. There is no reason to attribute his dislike to childhood trauma. The dissonance between Indianness and naturalness in the natural philosophy system provides a more plausible and dignified "explanation" of Bartram's mature character.

John received only a sketchy formal education in a country

school but, according to his son, showed an "early inclination to the study of physic and surgery," a tendency which would naturally enough lead him to the careful study of plants, important as they were to eighteenth-century medicine. He told Crèvecoeur, and it is generally accepted as true, that as an adult he bought himself a Latin grammar, "applied to a neighboring schoolmaster" for instruction and learned enough in three months "to understand Linnaeus, which I purchased afterward." He "botanized" around the farm, and further improved his education by borrowing books from James Logan, who helped him with his Latin and taught him to use a microscope. He thus entered the intellectual community of Philadelphia and by 1744 became a member of the American Philosophical Society.

It was during these years that he drifted from his early Quakerism into deism. The Darby Meeting struggled with and finally disowned him for his "dark notions" and "opinions" that Christ was not the son of God, but Bartram was an independent man and ignored their discipline, continuing to sit in meeting when he chose. He had altered his allegiance and become one of the enlightened. Crèvecoeur reports that over his greenhouse door Bartram had inscribed Pope's lines, "Slave to no sect, who takes no private road,/But looks through nature, up to nature's God!"

Bartram was a thoroughly practical man of great energy and restless intellect. He supported his family of eleven children by farming the land left him by his uncle Isaac. He constantly improved it by draining, ditching and fertilizing, and protected it by rotating his crops. While such progressive farming techniques were then fashionable on both sides of the Atlantic, they were by no means universally practiced by small family-farmers like Bartram. Unlike some of the bourgeois and aristocratic virtuosi, Bartram retained throughout his life the farmer's vital interest in the quality of soil and in the potential of streams for power and irrigation.

This practicality distresses some of his more romantic biographers but constitutes, in part, the special "American" tone in much of his work. Like many New England and Pennsylvania farmers, he became as familiar with rocks as with soil; unlike many he also mastered stonecutting and masonry. He built four stone houses in his lifetime for which he carved even the stone-work around the windows and doors. An appealing revelation of the pride he took in this skill is contained in a letter (24 January 1757) to Jared Eliot, a fellow natural philosopher from Connecticut with whom he shared an interest in agricultural improvements (Darlington, *Memorials,* hereafter cited as M, 374–75). Bartram explains at some length his method of split-ting rocks (for door-sills, steps, window-cases, pig-troughs, water-troughs), a method that if applied properly could split rocks "as true, almost, as sawn timber." It was during these years of practical labor that Bartram was recruited as a nature reporter and began his travels in search of specimens, first in the imme-diate neighborhood, later as far as Pittsburgh, Onondaga, or Florida. His fellow Pennsylvanian, Joseph Breintnall, intro-duced him to the English Quaker merchant Peter Collinson and through him to the virtuosi of England and the continent (Stearns, p. 577). Bartram eventually made botanizing pay, and that was important, but at first it required effort over and above what was necessary to farm profitably. John did not share his son William's distaste for work and disinclination to attend to any but natural history interests.

Between the early 1730s and his death in 1777, Bartram rose from a successful provincial farmer with an interest in native plants to an internationally famous nature reporter. Any extended biography records in detail what this portrait can merely allude to—his successes, his trials, his work in that role. Nevertheless, the best source for the day-to-day reality of Bar-tram's work is William Darlington's 585 pages of correspon-dence in *Memorials of John Bartram and Humphry Marshall* (1849). In it are examples of his correspondence with Benjamin Franklin, Dr. Garden, William Byrd, Jared Eliot, Cadwallader Colden and others in American natural philosophy; pieces of his correspondence with Linnaeus, Dillenius, Gronovius, Sir Hans Sloane, Dr. John Fothergill, Dr. Lettsom, Peter Collinson and others from abroad. This correspondence contains allusions to and versions of the items he sent to the *Philosophical Trans-actions* of the Royal Society of London over a period of more than twenty years. It also preserves engaging accounts of the hazards of travel and the difficulties of collecting seeds and shipping specimens to England.

Much of John's collecting was done in his immediate neighborhood, for like Thoreau he became intimate with the birds, herbs, trees, stones and animals that occupied the wet and dry lands around his farm. The Darlington collection suggests both the sweep and particularity of Bartram's interests: leaves, pine cones, live trees, seeds, hummingbirds' nests, eggs, live snapping turtles, snakes, frogs, butterflies, wasps, cicadae, hornets' nests, ores, fossils, persimmons, ginseng, etc. He was no mere seed factor nor merely a collector but an imaginative reporter as well. His interests extended from sexuality in plants to the geological origins of limestone, from persimmon liquor to a plan for a geological survey of America, from the "fibrous roots" of the salt-water mussel to aurora borealis, and from rattlesnakes' "small teeth" to fossil remains of mammoths on the Ohio.[3]

Eventually Bartram travelled far from home in his search for typical and exotic instances of nature's plenitude. From his thirty-ninth to his sixty-sixth year he made long, difficult and dangerous journeys through the wilderness. As early as 1736 he searched for the source of the Schuylkill River. By 1738 he had travelled through Maryland and Virginia to Williamsburg and west over the Blue Ridge Mountains. By 1742 he had extended his travels to the Mohawk River and Catskill Mountains, and in 1743 he travelled north to the Five Nations and Lake Ontario, sending his *Observations on the Inhabitants, Climate, Soil, Rivers, Productions, Animals, and other matters worthy of Notice ... from Pensilvania to Onondago, Oswego and the Lake Ontario* (hereafter cited as o) to Collinson in England. He travelled south through the Carolinas in 1760 and west to newly won Pittsburgh in 1761 (keeping a journal which is unfortunately lost), and through the Carolinas again in 1762. In 1765, as King's Botanist, he travelled through Georgia and East Florida with his son and wrote a journal, portions of which appear as an addition to William Stork's *An Account of East Florida* (1767).[4]

Bartram achieved monumental success in his peculiar calling. Two conditions especially combined to favor this success: he was at the right place at the right time, and he had made himself over into the right man for the job. As rationalization of the heavens had engaged the seventeenth century, rationalization of flora and fauna engaged the Linnaeuses and Gronoviuses of the eighteenth, and they particularly wanted specimens from the new world. Bartram was one of many to answer the need. His correspondence reveals the tone and mechanics of this "imperial" patronage of colonial field research, his essentially provincial status as he filled orders for plants from his English and continental patrons. Those virtuosi, one gathers, were to do the theorizing and categorizing; Americans, the collecting of raw materials. Even his friend Peter Collinson occasionally assumed the air of a fatherly teacher to this man only five years his junior. Bartram occasionally chafed under such patronage and under the colonial status assigned him by the English, but was usually too interested in his work, too engaged in his mission, to bother. He was a botanist in the new world, a respected member of the virtuosi of the largest city in the colonies, and a valued arm of the international community of natural philosophy. He was a product of his age, of his location and of current intellectual concerns. That he was more—a more memorable contributor finally than Dr. Garden, William Young or Dr. Witt—must be attributed not to general conditions but to his personal qualities of perseverance, perception, sensitivity and imagination.[5]

In part because of his success, Bartram's prose has failed to attract the serious and sustained scrutiny it merits. The causes of this selective neglect are not hard to reconstruct. It is no accident, no reprehensible oversight of scholars, but a deliberate neglect; for in one sense Bartram failed as a writer, a victim of the system to which he subscribed and which discouraged, or at least failed to nurture, elements of his native talent, humor, spirit. He wrote no popular book, and the quality of his expression stirred neither the public of his own day nor the scholars of a later age. Even Moses Coit Tyler who could find "strokes of poetic beauty" and "serene eloquence" in the spare prose of Professor Winthrop says of Bartram, whom he admired, that his books had "no literary merit besides simplicity and directness of statement" (AHAL, 521–23). Though not entirely supportable in particular instances, Tyler's judgment is generally a fair one from a literary point of view.

Bartram sought and achieved his prominence in natural philosophy without deference to literary considerations, depending less on the quality of his prose than on the quality of the packages and information he sent. He lacked Carver's drive to entertain as well as instruct and wrote almost exclusively for the fraternity of naturalists. He only reluctantly authored two "travels" and in them rigorously pursued a plain style, believing that facts, not art, were his proper concern. Even so his works contain instances of vivid description, humor, simple beauty and serene eloquence. His letters especially contain passages of un-self-conscious joy, tart and ironic humor, playful wit and some descriptions that are at once so vivid and economical in expression as to impress the reader with imagistic force.

How is it then that Bartram's prose has been so little scrutinized by students of American culture? Uneven literary quality is not sufficient to account for the neglect: anthologists have quite comfortably edited others of similarly modest literary accomplishment when the cultural content of their work seemed to justify the effort. A hint may be contained in the way that John Bartram and his son have been repeatedly yoked by historians, biographers and critics. *The Cambridge History of American Literature* found it "difficult to mention the son without reference to the father" (pp. 194–95) and Ernest Earnest (*John and William Bartram*, 1940) treated them as a biographical unit, as if John's promise were fulfilled in the accomplishments of his son. John appears in the literary histories primarily as the subject of Crèvecoeur's portrait in "Letter XI" of *Letters from an American Farmer*, and as father of William (1739–1823), as though his chief functions in American culture had been to generate raw material for Crèvecoeur's art and to father the man whose *Travels* (1791) fed the romanticism of Coleridge, Wordsworth, Chateaubriand and others. The type, not the whole man or his writing, captured the imagination of Americans. Crèvecoeur characterized him as the "enlightened botanist ... who united all the simplicity of rustic manners to the most useful learning," a simple, earnest, self-taught Pennsylvania botanist of firm social ideals, plain speech, and elevated sentiments. One day while plowing, according to Crèvecoeur, Bartram plucked a daisy and thought to himself, "What a shame ... that thee shouldest have employed so many years in tilling the earth and destroying so many flowers and plants, without being acquainted with their structures and their uses!" Moses Coit Tyler found none of the early American students of nature "pleasanter to recall than the Quaker naturalist, John Bartram." Recalling Bartram's being orphaned at thirteen, his slight schooling and his love of nature, Tyler celebrates his "high attainments in science" in spite of "outward disadvantages." While he sees no literary merit in Bartram's books, he does find them "interesting and good as the jottings of an eager naturalist while passing through a new world." [6]

Perhaps Bartram's prose has failed to engage the sympathetic interest of cultural historians partly because there seemed no need. His son was writer enough for both, and taken together, father and son illustrate certain themes of sectional and national cultural history that appeal to the mythic imagination. As the Winthrop family illustrates the shifting patterns of New England culture from the early-seventeenth-century medievalism of Governor Winthrop to the early-eighteenth-century enlightenment of Professor John Winthrop, the Bartrams seem to link the worlds of Benjamin Franklin and Washington Irving. From John, the simple, deistic, handy, plain, self-educated botanist, comes William, the romantic, sensitive, rebellious traveller-artist. Their family story seems to justify (as that of Benjamin Franklin and his son William does not) the American cultural experiment; John's simple, agrarian, colonial potential finding fulfillment in William's refined sentiment, gentler ethos and lofty aspirations.

Frederick B. Tolles's treatment of Bartram in the *Literary History of the United States* (hereafter cited as LHUS) displays the standard literary-history assessment of Bartram's prose. Tolles characterizes Bartram as a "simple Philadelphia Quaker turned deist" and says of his writing:

The laconic style in which Bartram set down his observations was a product of the Quaker tradition of plain speech reinforced by the spirit of scientific accuracy and objectivity. Traveling through the verdant Susquehanna valley in the summer of 1743, he allowed himself to note only that 'the land hereabouts is middling white oak and huckleberry land. . . . we went up a vale of middling soil, covered with high oak Timber, nearly west to the top of the hill . . . from whence we had a fair prospect of the river Susquehanah.' Even in the midst of the gorgeous exotic scenery of Florida, the taciturn Quaker naturalist, distrustful of emotion, permitted himself only a restrained scientific curiosity. [P. 91]

But Bartram was more than a simple Quaker, and his prose is often better than that Tolles describes—certainly more interesting as a cultural document. Much of Bartram's prose is less "restrained" and emotionally barren than these fragments taken from his *Observations* suggest. In his letters to Collinson and others, Bartram reveals considerable humanity as well as a facility for sharp and particular description. While the *Observations* is less rewarding in these respects, it conveys Bartram's joy while travelling, his irritation with mosquitoes, and his amused surprise at Iroquois "false face" customs. Indeed, even in the pages immediately preceding the section quoted by Tolles, Bartram describes an encounter with a rattlesnake which demonstrates that his reportage can be terse, vivid, simple and active:

After dinner . . . we ascended a great stoney ridge about a mile steep, and terribly stoney most of the way: near the top is a fine tho' small spring of good water. At this place we were warned by the well known alarm to keep our distance from an enraged rattle snake that had put himself into a coiled posture of defence, within a dozen yards of our path, but we punished his rage by striking him dead on the spot: He had been highly irritated by an *Indian* dog that barked eagerly at him, but was cunning enough to keep out of his reach, or nimble enough to avoid the snake when he sprung at him. We took notice that while provoked, he contracted the muscles of his scales so as to appear very bright and shining, but after the mortal stroke, his splendor was much diminished, this is likewise the case of many of our snakes.[7]

Many things may be said of this passage. Laconic it is; devoid of emotional thrust it is not. Action is bracketed by a description of terrain at the opening and a generalization on snake behavior at the end. The relaxed tone of this frame emphasizes the rapid challenge and response of the encounter itself. Bartram tells it as fast as it happens and even achieves a sound that echoes the sense in "we punished his rage by striking him dead on the spot." Bartram's diction communicates an urgency that belies mere "restrained scientific curiosity": the adjectives *enraged, irritated, cunning* and *nimble* and the active verbs *punished, barked, sprung* and *contracted* reveal the involvement of the author and convey the excitement of the meeting well beyond what one would expect in a bare "statement of the facts." One might even detect a note of subdued humor and self-satisfaction in the understatement, "his splendor was much diminished . . . after the mortal stroke." [8]

Though Bartram's reportage, especially his letters, will yield a number of remarkable fragments, to focus on these would merely reverse Tolles's perspective and give an equally distorted view of Bartram's accomplishment, intentions, sensibilities. Bartram authored no lost masterpieces that deserve a place on the shelf next to *Letters from an American Farmer,* Carver's *Travels, Notes on Virginia,* or William Bartram's *Travels.* As Tolles realized, Bartram sought mainly to record the simple truth in letters and journals directed to other virtuosi. He expressed it by minutely and accurately describing appearances and behavior. He rather distrusted art as he understood it and was frequently more careless in matters of style, form and expression than were his contemporaries who reported on nature. Nevertheless, his prose often expresses more than he himself may have realized; certainly more than critics and historians have allowed.

The common scholarly evaluation of Bartram's prose suggests that traditional disciplinary scholars subscribe to models of excellence that make it difficult to appreciate Bartram's written contribution to American culture. To the literary historian Bartram's prose is not art; to the historians of science his tone, expression, form and figurative language are irrelevant. Bartram the whole man is lost somewhere between science and literature. He survives into the twentieth century as a partial man only, a victim of disciplinary Balkanization of culture. Total alertness is required if we are to catch Bartram alive. We must doubt all customary assessments of the man, doubt even his own avowals of belief and value unless they suit *all* the evidence. Bartram has a way of surprising the reader whose guard is down. One illustration should clarify the danger.

Ernest Earnest, in his excellent study of father and son, says that "John's meager knowledge of the very things taught in the early schools was a sensitive point with him, as is revealed in a letter to Peter Collinson, who had apparently criticized him on this score." Earnest cites as evidence of this "sensitivity" one sentence from a letter of November 3, 1754: "Good grammar and good spelling, may please those that are more taken with a fine superficial flourish than real truth; but my chief aim was to inform my readers of the true, real, distinguishing character of each genus, and where and how, each species differed from one another, of the same genus" (M, 196). What Earnest fails to take into account is the ubiquity of such conventional "apologies" as this. It is more generic than ingenuous here and contains at least as much pride as humility. Bartram, himself, continues the letter with the challenge, "if you find that my descriptions are not agreeable with the specimens, pray let me know where the disagreement is" (M, 196). This from the man who six weeks later said in praise of his son's drawings that they were "not according to grammar rules, or science, but nature," suggests that he adheres intentionally to standards in which grammar and spelling are inconsequential compared to truth (M, 197). Bartram is not voicing his humiliation in the fragment quoted by Earnest. On the contrary, he is chiding Collinson for mistaking the essentials of truthful reports.

Dr. Samuel Johnson recognized only one requirement for the pursuit of botany when he said, "should I wish to become a botanist, I must first turn myself into a reptile." To "number the streaks of the tulip" one had to look closely and intensely at objects. Implicit in Johnson's quip is derogation, certainly, of the botanist's restricted perspective, as well as allusion to his own weak eyesight. But John Bartram was not so restricted. He saw character in wasps, volition in tipitiwitchets, excitement in pine trees, order in bird migrations and glory in the heavens. His eager search for knowledge, his tenderness toward animals, his humor and energy, his respect for facts and his disdain of superstition combined to make Bartram a nature reporter without equal. He lived and worked at the cutting edge of eighteenth-century natural history. This made him as great as he was; kept him perhaps from becoming greater.[9]

Persimmons Raw, and Thorough Ripe

How then should one approach John Bartram's writings? His written work exhibits a kind of "grammar" linking nature to culture, personality to Other, raw data to meaning. In his best pieces, the elements converge and nearly dissolve in the whole, leaving a configuration in which all interact to communicate exactly the content and tone of what Bartram knows and feels. In these cases even the structure and style "comment." The structure of his "grammar" appears most starkly, however, in his unpolished pieces, which also have the advantage of being wholly free from friendly "correction" and "improvement" by others. The study of his writing should start with this crude prose. One excerpt from his "Diary of a Journey Through the Carolinas, Georgia, and Florida" (1765–66, hereafter cited as D) serves to establish the most basic level of his composition:

dined at Mark hoggins a good kind man: therm. 91. here it had not rained to wet ye ground deep since January. these two days I found several curious species of plants in ye savanas trees was pine evergreen oak broad leaved willow scarlet scrubby white & red & black hicory . . . ash fartile berry andomeda Liquid amber all over Cornus Saw several curious plants & much tall piney ground then crossed A ferry to georgetown containing 150 houses pretily scituated on ye river sampets. . . . loged at Pikes on ye west end of long bay lightned all night. [D, 14]

Compositionally, this "Diary" occupies a position about a third of the way along a continuum from (Carver's) survey log to manuscript "Journal." It contains numerous terse entries such as "december 1 rainny morning therm 68 cleared up went to church PM showery & thunder" and "december ye 2 cloudy morning therm 70 A showery day PM 79" (D, 36).

Nothing could be much plainer, less self-conscious, more purely factual. Yet even in *this* "Diary" passages occur which demonstrate Bartram's reliance on homely figures and vernacular experience to convey accurately the look and feel of things:

after breakfast Mr Galpin took 5 or 6 hands & two battoes & provisions down ye river 12 miles to ye lovliest spring bank above 150 foot high at ye bottom of which issued out ye lovliest spring of clear water I saw in all this Journey it cam out in A body about as much as would run through ye bung hole of A hogshead very sweet & Just cold enough to drink heartily off about 6 foot above ye surface of ye river: this perpendicular bank is formed of oister shels all broke to small particles like cours sand & indurated so as to support A perpendicular position about as hard as ye bermudous rocks used for building & Just such A consistency one may cut it very easily with A knife. [D, 25]

This description has the appearance of one written extemporaneously and with little attention to style; yet its direct syntax and rapid rhythm, together with a vivid simile and the taste of cold water, show that Bartram, like Carver, could marshal words to describe a scene. Through all the surface crudities of the "Diary" shine still the qualities that are praiseworthy in Bartram's published writings. This is important to establish from his unedited "Diary" since it is also clear that Bartram's *Observations,* his contributions to the Royal Society *Philosophical Transactions,* and even his letters in Darlington's *Memorial* must all have been edited. Without the "Diary" one might be tempted, as critics have been with Carver, to credit editors with whatever felicities exist in the published work. In the "Diary," Bartram wrote on several levels: at times he merely jotted notes; at others he composed more extended and polished pieces, suggesting that when he wanted to he could write nearly standard English. It should be remembered that failure to punctuate is not nearly so distracting when the prose is handwritten and spaces can function to clarify pauses and stops.

Even at his most matter-of-fact, Bartram sometimes produced amusing combinations. The third day out by boat from Philadelphia (July 3, 1765) Bartram records that while "extream sick & head very dizey: three mile from shore 7 fathoms," he "saw A whale spout twice but could not bear lookeing up on deck" (D, 13). I take this juxtaposition of nausea and spouting to be deliberate and wry, but do not press the interpretation. Still Bartram shows elsewhere that he does adopt intentionally inappropriate expressions to achieve restrained ironic comment. For August 29, 1765, he includes a nearly four hundred word narrative—lengthy for the "Diary"—of his frustrations with southern hospitality. Three plantation owners in succession were so "ill natured" as to refuse him lodging. The third relented only after Bartram threatened to lie down in his field. Bartram and his son Billy were allowed to occupy an outbuilding worse than the Negroes' own, and Bartram exasperatedly concludes his narrative with a description of their lodging:

yᵉ little ould corn house was cleaned by pushing yᵉ ould mouldy corn in A heap on one side to make room for An ould bed to be laid on yᵉ floor for us to rest upon   when yᵉ gentleman tould us our room was ready for us & that we might go there as soon as we pleased   so we went directly about 100 yards from yᵉ dwelling hous   we found A candle set in an ould bottle for A candle stick & some hommony & two horn spoons to eat with   we pulled of our wett cloath & hung them up to dry & lay down amongst yᵉ ratts weevils grubs & muschatoes to refresh our selves after A tedious ride of near [35?] miles. [D, 21; brackets by Harper]

Certainly *refresh* is intentionally ironic. Furthermore, the exacting detail of the description indicates Bartram's dark enjoyment in recording completely the indignity he suffered at this plantation.

The variation between the terse log entries and the longer narratives and descriptions in the "Diary" shows that Bartram wrote sometimes merely to record facts and remind himself of what he had seen, and sometimes wrote descriptions that required only the slightest editing before being published. Bartram's best efforts do not appear in those portions of the "Diary" that have escaped editing. It is fair to assume, however, that even the best letters in the Darlington volume are corrected very little, probably only the punctuation, spelling and capitalization. The expression, imagery, rhythms, texture and tone are his own, as are the nonstandard grammar and informal diction. A most remarkable description in the letters is Bartram's description of the snapping turtle:

They are very large—of a dark muddy colour—large round tail, and feet with claws,—and old ones mossy on the back, and often several horse leeches sucking the superflous blood; a large head, sharp nose, and mouth wide enough to cram one's fist in,—very sharp gums, or lips, which you will,—with which they will catch hold of a stick, offered to them—or, if you had rather, your finger—which they will hold so fast that you may lift the turtle by it as high as your head, if you have strength or courage enough to lift them up so high by it. But as for their barking, I believe thy relator *barked,* instead of the turtle. They creep all over, in the mud, where they lie *perdu*; and when a duck, or fish, swims near them, they dart out their heads as quick as light, and snap him up. Their eggs are round as a bullet, and choice eating. [M, 196–97]

The reader is justified in inferring that the punctuation of this description has been edited, but the active verbs *cram, creep, dart* and *snap,* the rhythm of "they dart out their heads as quick as light and snap them up," the homely and accurate simile "round as a bullet," the detail of "horse leeches," the play on *bark,* and the sly humor of suggesting a finger be offered a snapping turtle are all Bartram's own. The humor even suggests something of the straight-faced extravagance put over on the "greenhorn"—something of the bland understatement so frequently characteristic of Mark Twain in *Roughing It*.

One of the surprises that awaits the student of Bartram is the discovery that he who described the snapping turtle with such economy and vernacular color could also pen, in a letter to Collinson, "O Pennsylvania! thou that was the most flourishing and peaceable province in North America, art now scoured by the most barbarous creatures in the universe. All ages, sexes, and stations, have no mercy extended to them." This attempt at high style results from his anger at the Indians who, he continues to say, "skip from tree to tree like monkeys; if in the mountains, like wild goats, they leap from rock to rock.... They are like the Angel of Death" (M, 206). Here he consciously adopts what seems to him "style," but the style is a conventional *belles lettres* potpourri of personification, exclamation, derogatory simile and grand epithet. Discovered in Freneau, this would seem merely stilted and derivative; in Bartram, who had a talent for unaffected reportage, it is depressing.

That John Bartram, himself, considered such excursions into neoclassical style appropriate to high seriousness is suggested by his letter to Peter Collinson after hearing of the death of Collinson's wife: "How grievous is it, for one that is thus agreeable, to be torn from our hearts! Her dear sweet bosom is cold; her tender heart—the centre of mutual love—is motionless; her dear arms are no more extended to embrace her beloved; the partner of his cares, and sharer of his pleasures must no more sit down with her husband at his table" (M, 194). There is no doubt that Bartram is expressing sincere emotion, but his choice of inverted syntax, "How grievous is it," and his insistence in probing the detailed losses of heart, arms and companionship results in a sentimental display as embarrassing to read as the bathetic nineteenth-century mortuary verse parodied by Mark Twain for *The Adventures of Huckleberry Finn* ("Ode to Stephen Dowling Bots, Dec'd," ch. 17). We must conclude that John Bartram conceived of style as a frosting, as an effect to be laid on for special occasions but which one need not struggle to achieve when addressing plain gentlemen on the simple facts of nature.

Bartram's attitude toward style complicates the critic's work, for one must assume that Bartram's most effective descriptions are not consciously devised, or at least that Bartram did not self-consciously manipulate diction, imagery and rhythmic ingredients so as to achieve memorable descriptions. Yet affecting images and rhythms crop up repeatedly in his prose. He describes the Euonymous: "its berries make a fine appearance, in the fall. It twists about the trees, or poles, like hops" (M, 173). His "natural" periods would appear to be tetrametric. Even the syntax of the second sentence echoes the sense of the content as it three times twists about "the trees, or poles, like hops." Another time he tells Collinson, "Pray give Catesby one root of Pawpaw" (M, 173). In neither instance (and there are many more) ought one conclude that Bartram sought these rhythmic effects. But there they are, and they suggest that his "ear" half-consciously recognized some rhythmic norm, probably out of his own speech pattern.

An author approaches the writer-as-craftsman role more closely when he adopts figurative speech to express himself. Bartram's best similes and metaphors seem unself-conscious, the fruit of vernacular, not "stylish" rhetoric. To emphasize his ignorance of American mosses, he tells Mark Catesby, "Before Doctor Dillenius gave me hint of it, I took no particular notice of Mosses, but looked upon them as a cow looks at a pair of new barn doors" (M, 321). There is the humor of Robert Frost in that. In another letter, he says of persimmons: "They are extremely disagreeable to eat until they are thorough ripe and will fall with shaking the tree: then their pulp is delicious. But their skin, which is as thin as the finest paper, still retains an astringent bitterness: yet many of our country people are so greedy of them, that they swallow down skin, pulp, and seeds, all together" (M, 122). One forgets that this is a botanist writing and not some traveller like William Byrd or Madame Knight. His inclusion of "shaking the tree" makes nicely personal the test of ripeness, and the simile, "as the finest paper," contributes tactile and visual concreteness to the thinness of the skin. The comic eagerness of the "country people" helps make the persimmons' deliciousness convincing.

Bartram was, at times, inclined toward rather extended bits of wit. In a letter to Collinson who shared his disdain for "theosophical moonshine," he pretended to adopt their friend Dr. Witt's notions of spiritual sympathy:

Now, though oracles be ceased, and thee hath not the spirit of divination,—yet, according to our friend Doctor Witt, we friends that love one another sincerely, may, by an extraordinary spirit of sympathy, not only know each other's desires, but may have a spiritual conversation at great distances one from another. Now, if this be truly so,—if I love thee sincerely—and thy love and friendship be so to me—thee must have a spiritual feeling and sense of what particular sorts of things will give satisfaction; and doth not thy actions make it manifest? for, what I send to thee for, thee hath chosen of just such sorts and colours as I wanted. Nay, as my wife and I are one, so she is initiated into this spiritual union; for she saith that if she had been there herself, she could not have pleased her fancy better. [M, 174]

Here Bartram is playful. He jokingly persists in developing, as if true, that which they both disbelieve. His playfulness with *if-then* logic and his development of the "spiritual conversation" conceit reduces Dr. Witt's notions to absurdity while at the same time flattering Collinson's taste and rationalism. He shows a sophisticated playfulness that one expects to find in Franklin's letters, but that one is unprepared to find if he sees Bartram as only a simple, earnest Quaker botanist. But Bartram's most compelling work takes strength from his straightforward and simple impulse to report particulars, rather than from his occasional excursions into wit or "style."

In the deceptively simple report of particular nature all extraneous "style" drops away and the man and the object appear startlingly clear to the reader. One such instance of Bartram's prose is his description of the way in which the groundsell distributes its seeds:

One day when the sun shone bright a little after its meridian, my Billy was looking up at it when he discovered an innumerable quantity of downy motes floating in the air between him and the sun. He immediately called me out of my study to see what they were. They rose higher and lower as they were wafted to and fro in the air, some very high and progressive with a fine breeze, some lowered and fell into my garden where we observed every particular detatchment of down spread in four or five rays with a seed of the *Groundsell* in its centre. How far these were carried by that breeze can't be known;

but I think they must have come near two miles from a meadow to reach my garden. As these are annual plants they do but little harm in the country.[10] [M, 387]

This passage has several fine qualities, not the least of which is its combination of objective description and subjective involvement of the persona. But it demonstrates a larger accomplishment as well, the apparently artless wedding of accidental sequence and logical contraction from the general to the specific. The seeds begin high in the air and drop into Bartram's garden where a specimen may be scrutinized; Bartram's mind proceeds from "a quantity of downy motes" to an individual seed; the images move from the sun, to groups of seeds, to "particular . . . down," to "a seed" alone. Then, from the specific encounter Bartram's thoughts soar again into the clear air of speculation. One is forcibly reminded of Jonathan Edwards's account of the flying spider, so often praised for similar qualities. Its form is characteristic of Bartram's best descriptions, those "thorough ripe" persimmons which "fall with shaking the tree" of his prose. Such is the nature reporter's "lyric," corresponding in shape and tenor to those nature poems by Transcendentalists which yoke Spirit and Nature in the eye of the seer.

An example of the characteristic way Bartram elevates facts of nature to cultural significance is his description of migrating birds and his speculations about what he observes:

Many birds, in their migration, are observed to go in flocks,—as the geese, brants, pigeons, and blackbirds; others flutter and hop from tree to tree, or upon the ground, feeding backwards and forwards, interspersed so that their progressive movement is not commonly observed. Our blue, or rather ash-coloured, great herons, and the white ones, do not observe a direct progression, but follow the banks of rivers—sometimes flying from one side to the other, sometimes a little backwards, but generally northward, until all places be supplied sufficiently where there is conveniency of food; for when some arrive at a particular place, and find as many before them as can readily find food, some of them move forward and some stay behind. For all these wild creatures, of one species, generally seem of one community; and rather than quarrel, will move still a farther distance, where there is more plenty of food—like Abraham and Lot; but most of our domestic animals are more like their masters: every one contends for his own dunghill, and is for driving all off that come to encroach them. [M, 211–12]

Not only does Bartram accurately describe real birds in this selection, he also declares a correspondence between ethology and ethics and uses that similarity to chide Man. In this and similar examples he demonstrates a continuity between John Winthrop's use of mice and snakes as moral symbols and Emerson's use of bees and squirrels, or Thoreau's use of groundhogs and crab apple trees. Bartram approaches Emerson even closer when he posits a feeling or volition in plants:

When we nearly examine the various motions of plants and flowers, in their evening contraction and morning expansion, they seem to be operated upon by something superior to only heat and cold, or shade and sunshine; . . . and if we won't allow them real feeling, or what we call sense, it must be some action next degree inferior to it, for which we want a proper epithet, or the immediate finger of God, to whom be all glory and praise. [Earnest, 64–65]

This is fundamental culture work, connecting noösphere, homisphere, biosphere.

At times Bartram achieves an intensity almost poetic. His best portraits live up to Marianne Moore's plea for "imaginary gardens with real toads in them":

I saw several of these wasps flying about a heap of sandy loam: they settled on it, and very nimbly scratched away the sand with their fore feet, to find their nests, whilst they held a large fly under their wings with one of their other feet: they crept with it into the hole, that lead [sic] to the nest, and staid there about three minutes, when they came out. With their hind feet, they threw the sand so dexterously over the hole, as not to be discovered: then taking flight, soon returned with more flies, settled down, uncovered the hole, and entered in with their prey.

This extraordinary operation raised my curiosity to try to find the entrance, but the sand fell in so fast, that I was prevented, until by repeated essays I was so lucky as to find one. It was six inches in the ground, and at the farther end lay a large magot, near an inch long, thick as a small goose quill, with several flies near it, and the remains of many more. These flies are provided for the magot to feed on, before it changes into a nymph state: then it eats no more untill [sic] it attains to a perfect wasp.

The order of Providence is very remarkable, in prescribing the different ways and means for this tribe of insects to perpetuate their species, no doubt for good ends and purposes, with which we may not be well acquainted, but most likely, for the prey food of other animals.

One kind of wasp fabricates an oblong nest of paper-like composition full of cells for the harbour of its young, and hangs it on the branch of a tree. Some build nests of clay, and feed their young with spiders; others sustain them with large green grasshoppers: then there are those, that build combs on the ground (like ours in England) to nourish a numerous brood. [This letter was transmitted to the Royal Society by Collinson, hence the parentheses.]

But this yellowish wasp takes a different method, with great pains digging a hole in the ground, lays its egg, which soon turns to a magot, then catches flies to support it, until it comes to maturity.

The wisdom of Providence is admirable, by giving annually a check to this prolific brood of noxious insects, in permitting all the males to die, which are the most numerous of the family; only reserving a few impregnated females to each species, to continue their race to another year.

Whereas bees, whose labours are so beneficial to mankind, always survive the winter to raise new colonies. [RSPT 53 (1763): 37–38]

These "*Observations . . . by Mr.* John Bartram *. . . on the Yellowish Wasp,*" though communicated to the Royal Society by Collinson and containing his parenthetical comment, unmistakably bear the signature of Bartram's style, perception, imagery and imagination. The active verbs *settled, scratched, held, crept* and *threw* and the compressed action in the series of predicates, "returned . . . , settled down, uncovered the hole, and entered with their prey," characterize Bartram's prose at its best. Also characteristic of Bartram is his use of precise details that surprise and delight with their accuracy and relevance. The images of the wasp with a fly under its wing, of the wasp kicking sand over the hole with its "hind feet," and of the specific greenness of the grasshopper entombed for the maggots, create a wasp as fully realized as any "real toad" in American literature. Even the homely "goose quill" simile adds conviction to the portrait. Bartram, the wasp and God (Providence) are stitched together.

Bartram "frames" his wasp portrait and thus creates an "imaginary garden" for it to live in. The first person narrator immediately funnels the reader's attention into the intense scrutiny of the wasp and virtually evokes the reptilian posture Samuel Johnson ascribed to botanists. Bartram's inclusion of the nesting habits of other wasps "fills in the corners" of the picture and supplements the portrait; Audubon and Catesby placed similar details in the corners of their pictures, including sometimes a sketch of a foot or bill, at other times a bit of landscape or an additional animal or plant. Finally, Bartram's speculations on Providence determine the tone and mood of the portrait in the same way that the addition of a classical column or a silver teapot in a Copley portrait establishes the place of the sitter in the scheme of things. Bartram has as surely created a portrait of meticulous likeness and forceful character as ever Copley did and has "framed" and "hung" it in the nature wing of the gallery of American culture.

"Wretched" Book—Revealing Document

Bartram created provocative and instructive "notes"; had he produced a *magnum opus* these bits are the sort of thing that would later have been collected and published as the "notebooks" or "journals" of John Bartram and been studied by scholars for additional insights into his creative strategies. The mass of his published work remains, however, undigested— either by scholars or by Bartram himself. Unlike Thoreau, he never mined his notebooks for the ore that could be refined and beaten into a sustained composition of merit. His *Observations* is an incomplete representation of his talent. Nevertheless, because of its greater length it reveals certain concerns and inclinations not exemplified by Bartram's shorter pieces.

John Bartram's *Observations* is the record of a trip in 1743 from Philadelphia up the Schuylkill-Susquehanna route to upstate New York at Onondaga, a side trip down the Oswego River to Lake Erie and the return trip to Philadelphia. He accompanied Conrad Weiser, who was being sent to an Iroquois council at Onondaga. Consequently much of the record is devoted to Indian manners, meals, morals and superstitions. Along the way, however, Bartram indulged his curiosity in plants, weather, soil, geology and fauna, as the long title suggests. Though he made three copies of this journal to send abroad, "several accidents" prevented any copy from being printed before 1751. And when it did appear Collinson, who had hoped it would "add to John's reputation, was disgusted by the wretched job" that was done by the printers. It shows signs of having been negligently proofread, if at all; numerous comma splices and fragments make it difficult to follow at times; verbs appear to have been omitted, or to have been made to appear to have been omitted by the paragraphing. The book is nevertheless a fascinating specimen of the type.[11]

Its eight-page anonymous "Preface" is at once an advertisement, a history of the book and author, an apology for plainness, and a plea for English settlement of the interior. The seventy-page "Observations made by Mr. John Bartram" makes up the bulk of the book, but a fifteen-page letter from Peter Kalm describing "the Great Fall of Niagara" is appended as "a proper supplement." Three plates are included: one is a map of Oswego, crudely done; another, "A View of the Fall of Niagara" from *Gentleman's Magazine*; and the third, a fold-out map of "Pensilvania, New-Jersey, New York" by Lewis Evans, who accompanied Weiser and Bartram on the trip. The section entitled "Observations," however (barring printer's commas), appears to be pure Bartram. Doubtless this is what disappointed Collinson; he probably hoped an editor would "improve" the manuscript for publication, as was customary. What was disappointing to Collinson thus becomes, to the modern reader, an opportunity to study Bartram at his task of extended composition and to discover his ideas of composition, particularly in questions of form.

Bartram's primary formal device is the chronological sequence, from which he deviates less often than does Carver. Occasionally he departs from strict chronology to insert anecdotes that "the Indians told," relevant excerpts from "a most judicious writer," or speculation on natural history questions, like the problem of the "flood" (o, 27, 51, 39). At the end of the chronology he inserts five pages of argument and speculation on the origin of the Indians, but there is little evidence of any radical rearrangement of materials such as Carver made to improve his book. Most of Bartram's entries begin with specific travel details; for example, "17th Day, we crossed the neck to the east branch of *Susquehannah*, up which we travelled along a rich bottom of high grass and woods of a fine creek . . ." (o, 32). Many entries include only such material. Others include fragmentary allusions to gnats, mosquitoes, rattlesnakes or Indian travelling companions. Some entries expand to several hundred words when describing Indian customs or speculating about natural history. These show Bartram's most sustained efforts at composition in the *Observations* and apparently occur where he had leisure to "regulate his journal" and where they are suggested by the place or situation described. Interesting as many of these are, they represent the simplest level of composition, demanding little attention to the whole by the author.

One looks almost in vain for evidence that Bartram consciously sought the larger rhythms of expansion and contraction that Carver achieved. The muted opening and closing effects achieved seem the unconscious function of the anatomy of the journey and of Bartram's own sense of excitement at the outset, of relief at the end. The entry for July 6, the fourth day out, takes the reader to the top of a hill which seems to operate as a "gateway" to the wilderness; there the "irritated" rattlesnake challenges Bartram, initiating the reader into the dangers ahead. Similarly, the entry for August 16, three days from the end of the journey, ends with Bartram's party coming to a house where "we heartily congratulated ourselves on the enjoyment of good bread, butter and milk, in a comfortable house, and clean straw to sleep on, free from fleas" (o, 74). Emotionally this remark returns the reader to civilization. One hesitates, however, to classify these as formal devices of composition because of what that implies of conscious design. I conclude instead that Bartram did not think of his *Observations* as literature, but only as a written account of a trip to which he added such observations and speculations as might improve its worth for instruction and in which he included occasional bits of personal narrative to help convey the difficulties and joys of travel.

Though the book as a whole is rather monotonously linear and cluttered with minute observations that few can read with profit or delight, particular portions reveal an admirable precision of observation and clarity of language. Oddly enough these tend not to be the sections devoted to flora, but those describing rattlesnakes, Indians, travel by canoe or horseback, and nocturnal noises or sleeping accommodations. The plants and trees are generally only listed, though exceptional ones are described: "we stopped for this night at the foot of a great hill, cloathed with large *Magnolia*, 2 feet diameter and 100 feet high" (o, 67). Presumably the trees he saw daily were well known, and merely to list them fulfilled his notion of the purpose of his journal.

In contrast, his imagination often seems engaged by the details of travel *per se*. He achieves an almost idyllic mood in one brief description of an interlude on horseback: "4th This was a fine day, and our traveling cool, because shady, and the gooseberries being now ripened, we were every now and then tempted to break off a bough and divert ourselves with picking them, tho' on horseback" (o, 64). And he describes canoeing down the Oswego River:

... we went down the river a mile N. big enough to carry a large boat, if the trees fallen into it where [sic] but carried away, this brought us to the river from the *Cayuga* country, near 100 yards wide, very still, and so deep we could see no bottom, the land on both sides very rich and low to within a mile of the *Oneido* river, where the river began to run swift, and the bottom became visible, tho' at a good depth. . . . a mile farther we came to a rippling, which carried us with prodigious swiftness down the stream, soon after we encountered a second, and a mile farther a third, very rough. In about an hour by the sun, after many other ripplings, we found our selves at the great fall, the whole breadth of the river which is above 100 yards wide and is eight or ten feet perpendicular: here we hawled our canoe ashore, took out all our baggage, and carried it on our back a mile to a little town, of about four or five cabins. [o, 46–47]

Bartram's manuscript was probably nearly bare of punctuation. The printer's introduction of commas is less than helpful and makes the account more difficult to read than a literal transcription would have been, but overlooking this impediment, the reader finds Bartram swiftly and accurately describing what canoeing on a river is like. As noted in other of his descriptions, he even displays syntactical rhythms that express a sense of the rapid progress from one "rippling" to the next. The conscientious particularity about height and breadth of falls, the depth and clarity of the water, the speed of the stream, instead of interrupting the narrative, actually draw the reader in and induce alert attention.

About the only time one senses a self-conscious artifice in Bartram's *Observations* is when he indulges in wry comment. He says, for example, of one night's lodging, "It rained this night through our old, tho' newly erected lodging, which was an *Indian Cabin* that we took the liberty to remove, knowing they usually leave behind them a good stock [*flock*, 1895 ed.] of fleas on the ground they inhabit; however, the wet deprived me of my rest that I had taken so much pains to secure against the vermin" (o, 13–14). Usually he adds such personal details simply and unself-consciously, as when he says of one stream that "the water was chin deep most of the breadth, and so clear one might have seen a pin at the bottom" (o, 17). Here he seems not so much to be practicing style as to be reporting in the language natural to him the depth and clarity of the water.

Occasionally he creates what in a more consciously artistic author would be called a "set piece," as in his description of a "comical" Indian:

At night, soon after we were laid down to sleep, and our fire almost burnt out, we were entertained by a comical fellow, disguised in as odd a dress as *Indian* folly could invent; he had on a clumsy vizard of wood colour'd black, with a nose 4 or 5 inches long, a grinning mouth set awry, furnished with long teeth, round the eyes circles of bright brass, surrounded by a larger circle of white paint, from his forehead hung long tresses of buffaloes hair, and from the catch part of his head ropes made of the plated husks of *Indian* corn; I cannot recollect the whole of his dress, but that it was equally uncouth: he carried in one hand a long staff, in the other a calabash with small stones in it, for a rattle and this he rubbed up and down his staff; he would sometimes hold up his head and make a hideous noise like the braying of an ass. . . . I ask'd *Conrad Weiser* . . . what noise that was? and *Shickalamy* the *Indian* chief . . . called out, lye still *John*. I never heard him speak so much plain *English* before. The jack-pudding presently came up to us, and an *Indian* boy came with him and kindled our fire, that we might see his glittering eyes and antick postures as he hobbled round the fire, sometimes he would turn the Buffaloes hair on one side that we might take the better view of his ill-favoured phyz, when he had tired himself, which was sometime after he had well tired us, the boy that attended him struck 2 or 3 smart blows on the floor, at which the hobgoblin seemed surprised and . . . jumped fairly out of doors and disappeared. [o, 43–44]

Much could be said about this piece, about its language, its dramatic quality, its imagery. But the quality that most stands out is its tone. It is a piece of humor. It takes its humor from the mental posture of the deistic observer who watches with passive tolerance the slightly exasperating folly of a fellow human being. He shows his debt here to Addison, but like the American "whiggish" humorists of the nineteenth century, he combines accurate local color with a detached appraisal of the antics he encounters. The piece is remarkably different in tone from William Bartram's and Philip Freneau's more sentimental pieces about Indians.

This different tone makes one think again of the embarrassment of John Bartram's biographers when they try to explain his "failure" to appreciate the Indian as his son and as Peter Collinson did. The only explanation they can offer is that the death of John's father at the hands of Indians might have embittered him. They cannot tolerate this Hugh-Henry-Brackenridge tone of voice in their gentle botanist. If one recalls the difficulty Jonathan Carver had integrating Indians into nature, however, an alternative explanation suggests itself. These nature reporters emotionally distinguished between Indians and nature. They were often moved by and attentive to animals, streams, flowers and wonders (as indices of the plenitude, variety, harmony and balance of Providence) but viewed Indians as possibly heroic, occasionally treacherous, sometimes enviably free and often divertingly comic, but nevertheless as *human* beings who belonged as reporters did to the middle link of the Chain of Being and who deserved to be judged as fellow men. The composition of the *Observations* helps confirm this conclusion. The Indian material and the reports on soils, plants, animals and streams are separated physically, sometimes distantly—as in placing the remarks on the origin of the Indians at the end of his journal—sometimes separated merely within a day's entry. They tend to be separated by style and tone, as well. About nature's products Bartram is seldom humorous; about himself and his trials he is frequently ironic or wry; and about Indians he is apt to be tart, disgusted, amused and sometimes approving, but always his reaction implies civilized standards of judgment. Bartram might read lessons for man in the behavior of wasps and birds but not in the behavior of savages. Were Indians rational, they would read the same lessons from nature that Bartram did. The romantics would define a noble role for the Indian, but this nature reporter would not.

The *Observations* is a short book. It is less packed with the "real toads" and evocative descriptions of "imaginary gardens" than one might wish. It lacks the density of his letters to Collinson, on the one hand, and the rhetorical qualities appropriate to extended works, on the other. Bartram never understood the potential of the nature-reporter role, never fully appreciated the wider cultural dimensions of his mission: to serve the growing public demand for news of the cosmos of nature. Instead he perceived himself as a reporter to the initiated and expert students of nature, as if responsive to the evolving paradigmatic organization of biology rather than to the evolving nonparadigmatic, cultural reorganization of the cosmos. His humor, imagination, intellectual excitement and emotional engagement occasionally sparkle above the reportage, but they are seldom marshalled to transform this longer work into a broadly engaging piece of nature reportage.

Gilbert White and the "Interestedness" of Bartram

The conclusion that forces itself upon the reader who ponders what Bartram wrote is that Bartram could write well when fully engaged. He could create memorable portraits of individual specimens like the wasp and the snapping turtle; he could write active scenes like the canoe trip down the Oswego; he could create dramatic incidents like that of the "comic" Iroquois; he could be sardonic, playful and wry as well as speculative and philosophical; he could lie on his belly to observe a wasp or stand gazing in awe at the heavens. In short, he could have made his *Observations* or his "Diary" a milestone in American literature. That he did not reflects a kind of failure in mid-eighteenth-century American values, especially those of Franklin's circle and generally those of the international community of natural philosophy.

There was little to encourage Bartram to regulate his journals in the larger sense. The project to which he contributed explicitly valued facts, plain speech and economy. To New Englanders such as Jonathan Carver or Benjamin Franklin—raised in a subculture that surrounded one with examples of sustained pulpit oratory, that rewarded with publication sustained compositional logic and organization, that tolerated the "massy" prose of Cotton Mather—plain speech and economy became disciplines that helped one achieve clarity, precision and vitality in writing. But for Bartram, born into a Quaker subculture that scorned and distrusted pulpit eloquence and trained by men like Collinson and the Philadelphia Junto to seek plainness and economy alone, there was no nurturing the esthetic sense that would have impelled him to make of a travel journal any more than an honest daily record of facts and speculations. That even so he created striking portraits in prose must be attributed to his genius; that he did not create an eighteenth-century "Ktaadn" or "Cape Cod" (Thoreau) is the fault of his milieu, religious and intellectual. Had he written his "Cape Cod" instead of leaving only jottings, the tradition in American literature that stretches from Thomas Morton and John Smith to Henry David Thoreau would have been the richer.

To begin drawing conclusions about national or sectional cultures, suggests the advisability of taking a look at John Bartram from a wider perspective. One might gain such a perspective from studying any number of Bartram's foreign correspondents, but since questions of literary merit have been raised by others and comparisons between Bartram and Thoreau have been made by me, the example of Gilbert White provides the most useful contrast to Bartram. Gilbert White (1720–93) "numbered the streaks of the tulip" too; he sketched fossils, described bats, measured moose carcasses, speculated on the hibernation of swifts, studied the aurora borealis, described mouse nests, measured trees and recorded arrivals of birds and incidents of weather in his *Garden Kalendar* as early as 1751. He communicated his observations and speculations to the "distinguished naturalists, Thomas Pennant and the Hon. Daines Barrington, with whom from 1767 he carried on a correspondence which formed the basis of his" *Natural History and Antiquities of Selborne* (1789). In countless remarkable ways his style, content, form and imagery resemble the best of Bartram, but in other ways he is quite different. His style is consistently felicitous, as Bartram's is not. His tone evokes a distinctly English and settled ethos; Bartram's, a colonial and peripatetic one.[12]

Gilbert White was born in Selborne, England, and educated at Basingstoke by the father of Joseph and Thomas Warton. He attended Oxford, graduated in 1743, became a fellow in 1744, master of arts in 1746, and was made a "senior proctor" in 1752. Though academic careers were several times opened to him, he chose to retire to Selborne as curate and devote his life to recording nature in his native village and to reading, poetry and correspondence.

It is clear as soon as one examines the works of White that he used the same genres as Bartram. His *Garden Kalendar* jottings are much like those in Bartram's "Diary":

May 1. Bombylius minor appears.
May 2. White frost, sun, cold air.
May 3. Made the annual-bed for a large three-light frame with 3 loads of dung.

.   .   .   .   .   .

May 26. Much gossamer. The air is full of floating cotton from the willows. There are young lapwings in the forest. Female wasps about: they rasp particles of wood from sound posts & rails. . . . Hornets collect beech-wood.[13]

He shares the initial concern for facts and an active economy of language born surely of the same Royal-Society "esthetic." When he turns to more extended descriptions of creatures, as in his *Selborne,* he, like Bartram, combines economy of language with precise vision to form vividly wrought portraits:

I was much entertained last summer with a tame bat, which would take flies out of a person's hand. If you gave it any thing to eat, it brought it's [sic] wings round before the mouth, hovering and hiding it's head in the manner of birds of prey when they feed. The adroitness it shewed in shearing off the wings of flies, which were always rejected, was worthy of observation, and pleased me much. Insects seemed to be most acceptable, though it did not refuse raw flesh when offered: so that the notion, that bats go down chimneys and gnaw men's bacon, seems no improbable story. While I amused myself with this wonderful quadruped, I saw it several times confute the vulgar opinion, that bats when down on a flat surface cannot get on the wing again, by rising with great ease from the floor. It ran, I observed, with more dispatch than I was aware of; but in a most ridiculous and grotesque manner. [*Selborne* 1: 56–57]

One is reminded of Bartram's description of the "yellowish wasp" by details such as the "shearing off the wings" and by the persona's intimate involvement in the scene.

There are countless similarities between Bartram's and White's prose, but what one slowly becomes aware of when reading White's *Selborne* is that, for all the similarities between the two (and they are many), White's book is significantly different in tone. One senses a mild, pastoral contentment and pictures a settled, kindly English country curate, disinterestedly curious and delighted by the wonders of the "God of Nature" in his ancestral village (*Selborne* 1: 119). The American naturalist John Burroughs, in "Henry D. Thoreau," compares Thoreau and White in a way pertinent here:

Thoreau always sought to look through and beyond [Nature], and he missed seeing much there was in her.... I do not make this remark as a criticism, but to account for his failure to make any new or valuable contribution to natural history. He did not love Nature for her own sake, or the bird and the flower for their own sakes, or with an unmixed and disinterested love, as Gilbert White did, for instance, but for what he could make out of them.[14]

Burrough's remark about White is supported by the evidence. His evaluation of Thoreau's attitude toward nature, however, is even more interesting as it suggests not only the difference between White and Thoreau, but the difference between White and Bartram and the kinship of Thoreau and Bartram. Burroughs's "criticism" highlights a significant difference between Bartram's and White's attitudes as revealed in their prose. White was content and disinterested; Bartram was always looking through nature to significance. In his practical, Philosophical-Society mood Bartram sought out medicinal springs, streams that would turn mills, useful herbs, ways to split stones for building; in his speculative and essentially religious moods he sought lessons from wasps about Providence, lessons from rocks about the deluge, insights from plants about volition and feeling in the lower orders of creation. A practical contributor to natural philosophy he may have been, but always an interested rather than a disinterested one. One could say of his investigations of nature what Burroughs says of Thoreau's:

... he was always reaching after, and often grasping or inhaling. This is the mythical hound and horse and turtle-dove which he says in "Walden" he long ago lost.... He never abandons the search, and in every woodchuck-hole or muskrat den, in retreat of bird, or squirrel, or mouse, or fox that he pries into, in every walk and expedition to the fields or swamps or to the distant woods, in every spring note and call that he listens to so patiently, he hopes to get some clew to his lost treasures, to the effulgence that so provokingly eludes him.[15]

The difference between White and Bartram is that White writes natural history, Bartram, nature reportage. White stayed at home and let "nature's people" come to him; he had lost no hound, horse or turtle dove. Bartram searched at home for signs of the way they had gone, as did Thoreau, and was always drawn on to the interior where the tracks seemed to be leading, as was Thoreau, too. He searched south, north, at home and west as far as Pittsburgh, but the mystical quality of his search is further revealed by the power the Mississippi held over his imagination. Even in his *Observations,* a book about upstate New York, his imagination shifts to the Mississippi as a region that "cannot be paralleled in any other part of the world" (o, 51). As late as his sixty-fourth year the pull of the region still excited him:

O! what a noble discovery I could have made on the banks of the Ohio and Mississippi, if I had gone down, and the Indians had been peaceably inclined, as I knew many plants that grew on its northeast branches.... I read lately, in our newspaper, of a noble and absolutely necessary scheme that was proposed in England, if it was practicable; that was, to search all the country of Canada and Louisiana for all natural productions, convenient situations for manufactories, and different soils, minerals, and vegetable. The last of which I dare take upon myself, as I know more of the North American plants than any others. [M, 254]

Bartram knew American plants better than any other man and yet was dissatisfied. Like Jonathan Carver, he collected useful and marvelous information but was drawn on further by the stronger attraction of what he had not seen and of which he had only heard provocative reports. For Carver it was the Oregon and the "Shining Mountains"; for Bartram it was the Mississippi. He was not a disinterested naturalist any more than Thoreau; he was a seeker, but one whose flights of imagination were triggered by specific and concrete nature.

An ingredient of Bartram's travels that has irritated and perplexed biographers like Josephine Herbst (*New Green World*), who seems to wish that John had been more like his son, is John's tendency to gauge the depth and width of a stream and report that it "is deep enough for flat-bottomed boats" or "big enought to turn a mill"; that a spring "gushes out between the rocks ... in quantity sufficient to fill a pipe an inch square"; or of an island, that the land is "pretty rich ... near the river, but the higher end sandy, ... it therefore produces little but *pitch pine"* (0, 30, 19, 18). The best Josephine Herbst can say of these entries is that they show that Bartram could not "be deflected from his duty to make a true report to the Crown." But she clearly implies that these practical observations represent a cheapening of his talents, a kind of prostitution of his genius to the ethic of Poor Richard (Herbst, 33). The fact is that such social and economic usefulness in nature was just as eagerly appreciated by him as were the effulgences of nature. Bartram and Carver were not alienated renegades but surveyors for potential settlements in nature for men; not lone wolves so much as honey bees, discovering and leading the way for colonists who would follow.

The practical purpose and the mythic purpose are wed in their travels. To realize that allows one to discern emotional content in Bartram's travels where otherwise one would not have perceived it—in his laconic, daily record of the minutiae of places stopped at, streams crossed, places lodged in, transport employed. To the uninitiated these are the sort of raw and useless details of which Samuel Johnson complained. But seen rightly they signify an emotional record as surely as do the sensitive descriptions of pine trees and natural springs, for it is the fact as well as the significance of travelling that stirs Bartram's emotions. Many of his most memorable sentences are those that describe him canoeing, climbing, wading and stopping for the night. To be travelling is exhilarating and to record the progress, satisfying, even when the jottings are so bare and terse as to suggest only private notes—like the *x*'s on a tourist's road map. Bartram is like the automobile traveller of today who recounts to the bewilderment and boredom of relatives the travel details of his trip—the hour he left home, where he gassed up, the difficulties of bypassing Cleveland, the satisfying ease of turnpike driving, and the location of rest stops and motels. It is not that nothing important was seen on these days or that the traveller is too unimaginative to appreciate beauty and stop for it, but only that travelling itself, the movement toward a distant goal, is felt to be both emotionally satisfying and practical. After all, if travelling to and through nature as one seeks his lost hound is religiously, socially and personally significant, it follows that the road maps must be drawn so that others can follow. One is reminded of Thoreau's particular descriptions of planting beans, getting the wood for his cabin, and tallying the costs of lime, nails, boards, brick and chalk. He enjoys being a practical, capable traveller on the way to a better life.[16]

From its earliest days American culture has fostered the union of the practical and the supernatural: ideals are useful; things have spiritual significance. This union helps explain how Howells could comfortably embrace an *optimistic* realism and Thoreau, a *practical* idealism. Bartram is in the midst of this tradition. He sees spirit in a tipitiwitchet and mill-power in streams. John Winthrop reports, as if the symbolic truth were obvious, that mice ate through an edition of the common prayer book but spared the Greek Testament and the Psalms. Thoreau respects the woodchuck "as one of the natives" and thinks he "might learn some wisdom of him." Howells metaphorically directs writers to "the real grasshopper," the "simple, honest, and natural grasshopper" as a guide in art. John Steinbeck says that "it is advisable to look from the tide pool to the stars and then back to the tide pool again." [17]

There is perhaps no greater theme in American culture. As one who concentrated on minute description of nature rather than "effulgences," and as one who never fully worked his observations into a sustained piece of art, Bartram has been overlooked by historians of literature and his works dismissed as having "no literary merit besides simplicity and directness of statement." But if one reads Bartram with a fuller understanding he discovers not only reportorial triumphs and artistic limitations, but a fellow human presence lending continuity to American culture as it stretches from the seventeenth century to the nineteenth century, and the twentieth.

V

Mark Catesby,

A Georgian Reporter

The "Artist-Naturalist" and Nature Reportage

Though Jonathan Carver and John Bartram mainly used words to report nature, advances in printing had made graphic reportage (both woodcuts and engravings) increasingly common in eighteenth-century travel narratives and natural histories. One of the finest illustrated natural histories to be inspired by the impulse to number the streaks of the tulip was Mark Catesby's *The Natural History of Carolina . . .* (London, 1731–47). Catering both to the public's delight in natural beauty and to their thirst for information about the "natural productions of America," its 220 plates and accompanying text describe hundreds of American birds, snakes, insects, mammals, fish and plants. *Gentleman's Magazine* praised the book's "beauty, accuracy and splendor" (1752), and Richard Pulteney, late-eighteenth-century historian and biographer of botanists, called it "the most splendid of its kind that England . . . ever produced" (1790). Thoroughly appreciated by other virtuosi, *Carolina* earned its author Fellowship in the Royal Society of London, and was "the sole reference" for thirty-eight of the 100 nominal entries of North American birds included by Linnaeus in his 1748 edition of *Systema Naturae*. The book was equally admired and used as a standard reference by such knowledgeable Americans as John Bartram, Benjamin Franklin, and Thomas Jefferson.[1]

Unfortunately, this "most splendid" book is hardly known today except to a few connoisseurs and a relatively small group of scholars who pursue the history of science, and even they have not recognized it for the rich cultural document it is.

While *Carolina* expresses one man's personal delight in the beauty and utility of nature and demonstrates his heroic dedication to its accurate verbal and graphic portrayal, it also conveys, through its icons and prose, the aspirations, assumptions, cultural priorities and the totally human spirit that motivated many eighteenth-century nature reporters. The historical causes of Catesby's diminished reputation in the twentieth century are complex and varied and involve such factors as nationalism, the progress of science, and the revolutions in taste that have occurred since his death. But the most important factor in the failure of modern scholars to appreciate the full significance of Catesby's work is their inclination to approach his work with eyes shaded by traditional disciplinary blinders.

Catesby's case dramatically illustrates how structures of thought and expectation, developed to advance rational understanding and disciplined appreciation, may eventually inhibit awareness of meanings and values incarnate in documents that do not fit neatly into prevailing paradigms. If this was true of Carver's and Bartram's "travels to the interior," which violated the categorical lines between history and fiction or literature and natural philosophy, it is doubly true for Catesby and other illustrators of natural history whose work seems to straddle the chasm between fine arts and science as well as liquidate the boundaries between modern sciences like ornithology, botany, mammalogy. The gulf between science and art, especially, makes the task of appreciating Catesby's work extremely difficult within the conventions of present disciplinary scholarship.

One art historian, E. P. Richardson, partially bridged this gap between art and science by establishing a hybrid category called "artist-naturalists" for such men as John White, André Thevet, Jacques le Moyne de Morgues, William Bartram, Alexander Wilson, J. J. Audubon and M. J. Heade. He placed Catesby in this group and judged him the "first notable artist-naturalist to work in America." Coining the term "artist-naturalist" was a provocative and useful tactic of intradisciplinary reform and a service to students of American culture not bound to the disciplines of art history and criticism. But the device is finally too timid an invention to survive beyond its *ad hoc* utility for Richardson. It still excludes the verbal reportage that frequently is an integral part of the work of men like Catesby and Audubon, and it raises almost as many problems as it solves for one who attempts to expand its applicability. Should one, for instance, add Daniel Beard and Ernest Thompson Seton to the list, or Louis Agassiz Fuertes and Roger Tory Peterson? Or should the painters Morris Graves, Winslow Homer, George Catlin or Albert Bierstadt, or the photographer Eliot Porter be given partial membership on the basis of selected works? The amorphous category provides inadequate direction.

Another more serious problem is the way in which the artificial category "artist-naturalist" establishes standards of excellence and inferiority. That Audubon dominates the category unavoidably implies that others deserve categorical status in direct proportion as their work approximates his. Thus Catesby's initiating the convention of posing a bird on a relevant piece of flora, as Audubon later did, has earned him the praise of many. George F. Frick and Raymond P. Stearns, for example, titled their biography of Catesby, *Mark Catesby: The Colonial Audubon* (1961), as if to lend Audubon's fame to Catesby. In the "Preface" they advance as one proof of Catesby's worth that "his method of illustrating his subjects won him the posthumous compliment of imitation by no less a person than John James Audubon." While their claim is true and points to a fact worth knowing, their language reveals that they mistakenly accept Audubon as the standard by which Catesby's earlier art ought to be judged.

Worst of all, the hyphenated approach to interdisciplinary synthesis allows the dissimilar critical criteria of competing disciplines to remain distinct and antagonistic: the draftsmanship and composition of Catesby's art is judged with Audubon's romantic art as the standard of excellence; his delineation of fins, feathers, seeds and leaf veins is evaluated against the scientist's desire for structural accuracy. The inadequacy of the hyphenating approach is perhaps most dramatically exposed by pursuing it to its logical extreme; thus Catesby might accurately be called an artist-botanist-ichthyologist-herpetologist-ornithologist-entomologist-mammalogist-writer. Such an absurdity makes it apparent that a radically different alternative solution must be tried—one that both respects the coherence of the individual and his work and yet facilitates meaningful comparison and criticism by providing a perspective from which the student may inspect Catesby's work in relation to others of its kind. What is needed is a cultural approach that recognizes Catesby's resemblance to Jonathan Carver and John Bartram, men of his age and intentions, more than an approach which artificially dismembers Catesby's work and distributes the parts among competing disciplines.[2]

To group Catesby with Carver and Bartram as a nature reporter and to dissolve, thus, the barriers between graphic and verbal art may seem at first unwise, even reactionary, and as tending toward reestablishment of certain critical fallacies that were conscientiously cleansed from art studies by post-Victorian critics. But Catesby worked in an age that often spoke of graphic art as if it were an analogue of poetry and that wrote poetry as if the ideal were to achieve in words the visual impact of painting; consequently we do not violate the period's assumptions if prose writers, natural philosophers and "artist-naturalists" are grouped together. A stronger argument for dissolving the barriers in this case, however, is presented by the work itself. Catesby was one who recorded his "delight . . . in the visible framework of nature in the New World" by graphically and verbally numbering the streaks of the tulip. His work is not the grotesque offspring of an uneasy wedding of painting and prose, of art and science, but a piece unified by subject matter, feeling and purpose. His book is the product of a man who sought out nature and rendered it in words and line for the virtuoso and for the casual reader alike. To study Catesby as one who contributed to the body of nature reportage, a "literature" with detectable conventions and standards, will both increase our understanding of mid-eighteenth-century American culture and reveal that Catesby was an original and not just history's first try at the Audubon type.

"A Passion of Viewing"

Catesby's biography is presented in detail by Frick and Stearns. Yet certain items which illustrate his attitudes toward nature, explain his sense of mission, detail his travels and illuminate his methods of working deserve recitation here for they serve to advance a sense of the man's personality and experience and to establish his kinship with Carver, Bartram and other reporters.

Unlike Carver and Bartram, Catesby was favored with a family and circle of acquaintances qualified to nurture whatever early genius he may have shown in natural philosophy. Both sides of his family were educated and his mother's people included lawyers of antiquarian and "philosophic" avocation. They may well have introduced him to the study of nature in general and botany in particular. His father was a lawyer, a man of property, and a civic official of Sudbury in East Anglia whose known toleration of religious nonconformity during the Restoration may have opened for his son acquaintance with the nonconformist virtuoso Samuel Petto, that "Grave Divine" who, according to Frick and Stearns, contributed to the *Philosophical Transactions* (1699) of the Royal Society and corresponded with Increase Mather in America. While nothing substantial is known of Catesby's education, biographers conjecture that he may have been tutored by Petto; and Catesby's later "friend and fellow ornithological illustrator," George Edwards, claimed that Catesby fell into an acquaintance with the great John Ray and was inspired by that religious naturalist to pursue natural history. Catesby, himself, reports an "early Inclination . . . to search after Plants and other Productions in Nature," and his biographers have gathered considerable material to suggest that he spent his early manhood "amidst the eminent circle of his Uncle Jekyll's friends," who initiated him into botanical studies. Nevertheless, his most important experience was surely his first trip to America (1712–19), his "apprenticeship," as Frick and Stearns call it. He had "imbibed a passionate Desire of viewing," he tells us in the Preface to *Carolina,* "the Animal [and] Vegetable Productions in their Native Countries; which were Strangers to *England.*" His sister Elizabeth's residence in Virginia gave him the opportunity to satisfy that desire.[3]

Within a week of his arrival at Williamsburg (April 23, 1712) he had met William Byrd, F.R.S., and before long had become his friend and travelling companion. With Byrd he hunted hummingbirds' nests, "mended" the garden at Westover, travelled to Indian country at Pamunkey, shot a bear cub eating grapes in a tree, killed snakes and learned of the snakeroot's "virtue" as an antiveninous herb. He drank canary wine and met many of the Virginia establishment. He was generally introduced to colonial culture and to American nature. During these

years he collected and sent seeds and specimens to England and
explored the land from tidewater to Appalachians. Rather aim-
lessly he "gratified [his] Inclination in observing and admiring,"
he says, "the various Productions of those Countries." He made
some paintings of birds which were to help him secure on his
return to England the patronage he needed for his travels to
Carolina in 1722. These years of delight might be likened to
Jonathan Carver's experiences in the French and Indian Wars,
which changed the direction of his life, and to John Bartram's
early investigations of nature. Each thus "found" his vocation in
his middle years—in the early forties for Catesby and Bartram,
in the early fifties for Carver—and henceforth devoted all his
talent and energy, even health, to the pursuit of beauty and significance in nature.

Unless a professional man or independently wealthy, the
eighteenth-century natural philosopher needed patronage to
pursue extended studies of nature: Carver attached himself to
Robert Rogers; Bartram secured English patronage through
Peter Collinson; Catesby returned to England (1719) to gather
support for his projects. He was already known and his packets
of seeds and specimens appreciated, but he brought back paintings of "Birds etc." to show around, for he intended "againe to
returne" to America and sought encouragement from the virtuosi and financial support from patrons. He gained the support
of the Royal Society and of William Sherard (foremost English
botanist), the interest of Sir Hans Sloane (President of the Royal
College of Physicians and eventually President of the Royal
Society) and the favor of Governor Nicholson, soon bound for
his post in Charleston, South Carolina. It was 1722 before
Catesby was finally under way, but he went well furnished with
patrons and plans much as did John Bartram forty-three years
later.

His travels were, as were Bartram's, a mixture of delight,
frustration and hardship. He was so plagued by patrons who
apparently each expected a separate collection of drawings and
specimens that he wrote to Sherard, "I hope it can't be expected
I should Send Collections to every of my Subscribers, which is
impracticable for me to doe." He suffered illness, as would
Bartram later, and found similar difficulties in collecting, pack-
aging and shipping his specimens. Boxes for specimens had to
be made, and the price was "excessive." Birds had to be stuffed

with and packed in tobacco dust or preserved in rum, and even so, once they were shipped, they might be ruined by thirsty sailors who drank the rum in spite of the dead birds and lizards embalmed therein.

Nevertheless, he pursued his studies eagerly, trying always, as would William Bartram, to paint from nature exactly what he saw. Since colors of dead fish fade quickly, for example, he used several subjects, "having a succession of them procur'd while the former lost their colors." His biographers appear to regard this concern for vivid color as an extenuation but no real excuse for an occasional omission of a dorsal or pectoral fin. They are unable to evaluate his drawings by other than modern taxonomical standards. But Catesby was no disinterested drafts-man seeking only such details as would be useful to Linnaeus or later classifiers; he sought beauty as well as morphology. He thought that "Plants, and other Things done in a Flat, tho' exact manner, may serve the Purpose of Natural History, better in some Measure, than in a more bold and Painter-like Way," but his understanding of "the Purpose of Natural History" was different from present ideas. The purpose was complex and beauty constituted one of the essentials to be recorded, as his choice to paint bird "Cocks only, except two or three," shows: ". . . Males of the Feather'd Kind (except a very few) are more elegantly colour'd than the Females." "How lavishly Nature had Adorn'd" specimens with "Marks and Colours" was "most admirable" and surprising, he thought. John Bartram similarly regarded the "bright and shining . . . splendor" of the rattle-snake as significant enough to record (o, 12), and Carver re-peatedly described the colors of snakes beyond what was neces-sary for identification and added, "these are very Beautiful" (J1, 60–64).

In 1725 Catesby journeyed to the Bahamas to draw fish, but not before he had attempted to win support from England for a plan to travel to "the remoter parts of this Continent . . . in order to improve Natural Knowledge." Like Carver and Bartram, he thirsted for the distant interior, but when the plan failed he returned to England (1726) and spent the next twenty years engraving, tinting and supervising the publication of his *magnum opus*. There was no place in the colonies where quality engraving and printing could be done, even had he wished to print his book in America. Benjamin Franklin, for instance, was still in London improving his skills as a printer in 1726. Even a century later Audubon chose to go to England and Scotland to find engravers and printers worthy of producing his work.

Once in England, Catesby—like Carver—found himself both encouraged to publish his works and denied further financial support by the patrons who had sponsored the trip. He had hoped to have his drawings engraved in Amsterdam or Paris but was convinced by his friends that the expense would be too great. So he applied to Joseph Goupy for "Advice and Instructions" and began his second "apprenticeship": "I undertook, and was initiated in the way of Etching them myself." Similarly, Carver had bought himself surveying books, and Bartram had taught himself Latin. Unlike an adolescent apprentice, Catesby, by now in his mid-forties, had mature and vigorous opinions about representational accuracy, but having had no technical training to mold his sense of what was possible or conventional, he undertook what no professional engraver would have done —to abandon the "Graver-like manner . . . of Cross-Hatching, and to follow the humour of the Feathers, which is more laborious, and I hope . . . more to the purpose." He pioneered, also, in the "use of plants as backgrounds for most of his birds," an innovation of cognitive as well as esthetic significance. Probably also, as Frick and Stearns suggest, the innovation was an expedient way of including flora and fauna in his book without expensive multiplication of plates. Thus in his naïveté and enthusiasm, seventy-five years before Alexander Wilson and a century before Audubon, Catesby expressed in words, line and color, in a novel form and with a modified technique, his response to the "visible nature in the New World." In his attempt to wed beauty and representational accuracy of a quality demanded by natural philosophy, he created *The Natural History of Carolina,* a monumental addition to American nature reportage.

Success, honor and great work marked the last two decades of Catesby's life in London. From 1729 to 1747 he engraved and issued two hundred and twenty plates of birds, fish, snakes, plants, insects and mammals, first in sets of twenty so that the cost would not be prohibitive. The cost of the whole was to be twenty guineas. At first finances were a problem for Catesby, but then that amazing patron of colonial natural philosophers, Peter

Collinson, "lent him without interest" a "considerable sum of money" to prevent his work falling "a prey to the book sellers," as Bartram's and Carver's were later to do. He was made a Fellow of the Royal Society in 1733 and was a valued contributor to the Society's proceedings. He corresponded with English, continental and American natural philosophers and was sufficiently appreciated by the Linnaeans to be sought out by Peter Kalm for interview before that student of Linnaeus's left for America. Kalm's account of their meeting reveals a brand of humor in Catesby reminiscent of Franklin's in his tall tale of the whale leaping up Niagara Falls: Catesby warned Kalm that Virginia punch was so strong it paralyzed the fingers and forced one to hold his glass between his wrists to drink.[4]

"On April 16, 1747, Peter Collinson wrote Linnaeus, 'Catesby's noble work is finished,'" though the Appendix of twenty plates had not yet appeared. In that same year, at the age of sixty-four, Catesby clandestinely married Mrs. Elizabeth Rowland and at his death two years later, was survived by her and two children, Mark and Ann. (Whether these children, one of whom was eight, were Catesby's from a previous unrecorded marriage, were Elizabeth's, or were their "natural" offspring is not clear.) Catesby's marriage would seem to indicate that he did not intend to return to America. If he had ever intended to, it is nowhere recorded, though more than half his life and nearly all his adult life had been spent either travelling in America or executing and publishing his account of her natural "productions." His residence in England signifies little except that London was for him the "Center of all Science" and the center of English language publishing for the British Isles and America alike. Finally, of course, it is the sustained and artful negotiation between nature's "givens" and culture's "takens" that makes his *Natural History of Carolina* "a work which exceeds all of the same kind for its beauty, accuracy, and splendor" (*Gentleman's Magazine* 22:300) and a rich document of early American culture as well.

The Prose of Carolina

Catesby's prose has suffered even more neglect than his art. Except for his Preface, scholars have generally regarded his prose as inconsequential. As with Carver and Bartram, they have

looked *through* it more than *at* it and have taken it to be only
a vehicle of information, not a medium of expression worthy of
study in itself. Granted that the bulk of Catesby's prose functions mainly as a verbal complement to the plates, still some
of it is admirably sophisticated and subtle—in fact, more complex than Bartram's and more polished than Carver's. While
the iconography of *Carolina* is a "most elegant performance"
by itself, the book is far richer with the prose than without it.
To study *Carolina* as nature reportage in which verbal and
graphic art work together reveals its similarities to Carver's and
Bartram's work and clarifies the operation of Catesby's reportage: in spite of their considerably different size and pretension,
the folio *Carolina* shares numerous similarities of format, content, tone and texture with Bartram's tiny *Observations* and
Carver's octavo *Travels*. Recognition of such kinship provides
the necessary perspective for a fresh assessment of Catesby's important contribution to American culture.

   Carolina opens with a map of the territory, continues with
a seven-page "Preface," and expands into a forty-one page "An
Account of Carolina and the Bahama Islands." Only after these
fifty-one pages of folio size text does one arrive at the body of
the work, the 220 plates of illustrations and accompanying,
facing, textual remarks. The second volume concludes with an
"Appendix" of twenty more plates and text, just as Carver's
Travels ends with seventeen pages of appended remarks, and
Bartram's *Observations,* with the five-page discussion of Indian
origins and Peter Kalm's letter on Niagara Falls. The "Preface,"
typically for such works, reports the author's "early Inclinations" to study nature, explains his methods of collecting and
painting specimens and "apologizes" for his lack of conventional
skill but preference for plain truth in etching, color, prose.

   "An Account . . ." is a loosely organized collection of essays
and descriptions on various topics of interest to virtuosi. Like
part three of Carver's *Travels* it is only vaguely systematic,
grouping remarks under such divisional titles as "*Of the Air of
Carolina,*" "*Of the Soyl . . . ,*" "*Of the Water,*" "*Of the Abori-
gines of America,*" "*Of the Agriculture,*" "*Pinus,*" "*Of Beasts,*"
"*Some Remarks on American Birds,*" "*Of Insects,*" etc. Some
titles are subdivided; "*Of . . . Agriculture*" includes "*Of the
Grain . . . ,*" and within that topic, "*Indian Corn,*" "*Rice,*"

"Wheat." Under *"Pinus"* one can find a long article (as detailed and complete as Bartram's letter to Jared Eliot on stonecutting) on *"The Manner of making Tar and Pitch"*; under *"Of Fish,"* an explanation of how *"To make Caviar."* There is no attempt evident to achieve the inclusiveness or to create the discrete categories that one meets in nineteenth-century scientific works. He simply loosely groups together several remarks on phenomena he thinks will be interesting, useful or engaging, but in the process he informally articulates a rough ethnosemantic map of the "cosmos of nature." He reports wonders like the path left by a hurricane through the woods (ii) and the havoc left by a flood (vi–vii), speculates on the origins of the Indians (vii–viii), explains the geographical regions of the seaboard (iii–vi), or narrates encounters with bears (vi) and Indians (xiii).

The remainder of *Carolina* is primarily descriptive, pictorially and verbally. Volume One includes one hundred plates, largely of birds and plants, each faced by parallel English and French descriptions in normally plain, terse prose. An excerpt from the first volume reflects the prevailing tone: "The Bald Eagle ... weighs nine pounds: the Iris of the eye is white; ... the Tallons black, the Head and part of the Neck is white." He had said earlier, in the "Preface," that as pictures convey a "clearer Idea, ... I have been less prolix" in the text, "judging it unnecessary to tire the Reader with describing every Feather, yet I hope sufficient to distinguish them without Confusion." Volume Two includes one hundred plates of fish, crabs, tortoises, serpents, mammals, insects, birds and plants—marine and terrestrial; and the "Appendix" in Volume Two adds twenty plates of mixed subjects—partridges, wasps, beetles, fish, bison—all with facing pages of text. Only occasionally does the text reveal personal responses to nature or break into narrative, but the instances where it does are worth studying for their obvious similarity to fragments in Carver's and Bartram's works. These dilations occur most frequently in the last half of Volume Two and in the "Appendix." Perhaps as he began to see the end of his work in sight, Catesby strove to express more fully what he felt and knew of nature, and found language to be the more economical medium of the two he employed.

While the bulk of Catesby's textual descriptions are objective and terse, like that of the eagle, a few break the pattern.

Opposite plate forty-one (fig. 15) in which a rattlesnake is figured, Catesby recalls several incidents that neatly supplement in prose the restrained objectivity of the plate. Playing cleverly upon his readers' inevitable fascination and repulsion he balances generalizations about serpent anatomy and appearance with an account of his meeting with one specific snake. Beginning with its awesome dimensions and proceeding to its chilling behavior he effectively underscores the powerful, if restrained, threat of the snake pictured and creates a total impression of snakeness which neither words nor picture alone could do as well. "The largest I ever saw," he tells the reader, "was one about eight Feet in Length, weighing between eight and nine Pounds":

This Monster was gliding into the House of Colonel *Blake* of *Carolina;* and had certainly taken his [Abode] there undiscovered, had not the Domestick Animals alarmed the Family with their repeated Outcries; the Hogs, Dogs and Poultry united in their Hatred to him, shewing the greatest Consternation, by erecting their Bristles and Feathers, and expressing their Wrath and Indignation, surrounded him, but carefully kept their Distance; while he, regardless of their Threats, glided slowly along. [2: 41] [5]

Against the foil of his customarily laconic descriptions, this one stands out as emotionally vivid. He draws the reader into the scene, engages his emotions as well as his eye, and dramatizes the snake's threat to harmony by contrasting its cool disdain as it glides into a house with the domestic disorder it causes among the hogs, dogs and poultry whose anthropomorphic sentiments of wrath and indignation echo those of the human spectators. That he fully intended to play on the reader's fear and repulsion is illustrated by his report in the next paragraph that "it is not uncommon to have them come into Houses," or even into one's bed: "the Servant in making the Bed . . . (but a few Minutes after I left it)," he says, "discovered a Rattle-Snake, lying coiled between the Sheets." This peculiar American snake moved Catesby to abandon his usual plain exposition and to develop instead horrific anecdotes of surprise, superstition, rapid death and Indian remedies.

Catesby's verbal draftsmanship is as meticulous in rendering details as is his graphic. He describes in words, for example, the colors of the scarab beetle pictured in the Appendix (fig. 2):

The thorax of this is covered with a shield of a crimson metallick lustre, the head and lower part of the shield of the like lustre, blended with green. From the crown of the head rises a shining black horn recurved backward. The sheaths of the wings are ribbed, and of a shining deep green; as are the thighs and under-part of the abdomen. [App. 11]

Though somewhat more generalized at first, his description of the flotsam left by a flooded river is at last equally concrete:

Panthers, Bears and Deer, were drowned, and found lodg'd on the Limbs of trees. The smaller Animals suffered also in this Calamity; even Reptiles and Insects were dislodged from their Holes, and violently hurried away, and mixing with harder Substances were beat in Pieces, and their Fragments (after the Waters fell) were seen in many Places to cover the Ground. [An Account, vii]

While the occurrence is awesomely violent and dramatic, it is Catesby's use of relevant, specific items—shattered insects on the damp ground and animal carcasses in trees—that makes the description compelling and convincing.

Besides static, meticulous details, Catesby could convey movement. His account of the rattlesnake "gliding into the House of Colonel *Blake*" is one example. His account of dolphins in pursuit of flying fish is another. Not only does Catesby's prose report the action, but his syntax suspends the flying fish agonizingly in space for three adjectival phrases before dropping them exhausted into the jaws of the dolphins:

The Pursuit of Dolphins after Flying-Fish, was another Amusement we were often diverted with; the Dolphins having raised the Flying-Fish, by the swiftness of their Swimming, keep Pace with them, and pursue them so close that the Flying-Fish at length tired, and their Wings dry'd, and thereby necessitated to drop into the Water, often fall into the Jaws of their Pursuers. [Preface, vii]

In another description he shows a kind of baroque complexity of rhythms which suggests the verbal equivalent of the serpentine lines of his graphic designs in which snakes entwine themselves in plants (fig. 14):

The larger Rivers in *Carolina* and *Virginia* have their Sources in the *Apalatchian* [sic] Mountains, generally springing from Rocks, and forming Cascades and Waterfalls in various Manners, which being collected in their Course, and uniting into single Streams, cause abundance of narrow rapid Torrents, which falling into the lower Grounds, fill innumerable Brooks and Rivulets, all which contribute to form and supply the large Rivers. [An Account, vi]

The syntactical rhythms of his multiplying clauses do not so much halt the flow of the language as twist and divert it as it moves forward.

Catesby's imagery, of course, is best exemplified by his plates, but in his prose he occasionally creates precise portraits of active specimens, similar to Bartram's account of the "yellowish wasp." Interestingly, his account of the "Ichneumon Wasp" alludes to Collinson's account of similar wasps that Bartram sent from Pennsylvania (App. 4, 5, 13). Catesby meticulously describes the wasps, their clay nests, their way of crippling and entombing spiders for their nymphs' food, and their "surprising dexterity and odd gesticulations" as they sing their "odd musical notes" and labor to fabricate their nests (App. 5). His most compelling verbal portrait of this sort is of the scarab beetle (fig. 2), startlingly called by him the "Tumble-Turd." So vivid, entertaining, humorous and detailed is Catesby's description of the ludicrous industry of the beetle that one is tempted to perceive a satiric portrait of humanity in the tradition of Swift:

Their constant employ, in which they are indefatigable, is, in order to continue their Species, to provide proper nidi to deposite their eggs; this they do by forming round pellets of human dung, or that of cattle, in the middle of which they lay an egg.... I have attentively admir'd their industry, and mutual assisting one another in rolling these globular balls from the place they made them to that of their interment, which is usually the distance of some yards more or less; this they perform breech foremost, by raising their hind part, and forcing along the ball with their hind feet. Two or three are sometimes engaged in trundling one ball, which often meeting with impediments by the unevenness of the ground, is deserted by them, yet by others is again attempted with success; ... they and the balls are continually tumbling and rolling one over another down the little eminences; but not discouraged thereby, [they] repeat their attempts, and usually surmount these difficulties.

These Insects ... with the like sagacity of the *Turky Buzzard,* ... find out their subsistence by the excellency of their noses, which direct them in flights to the excrement just fallen from Man or Beast, on which they instantly drop, and fall unanimously to work in forming balls &c. which they temper with a mixture of earth. So intent they are at their work, that tho' handled, or otherwise interrupted, they persist in their oeconomical employment without apprehension of danger. [App. 11]

While some of the longer passages in *Carolina* bespeak Catesby's desire to communicate novel and useful intelligence, no such purpose excuses his extended treatment of this specimen. The dung beetle is not obviously useful to man and is not a specimen peculiar to the new world. The reason for his full description of the "Tumble-Turd" must be assumed to be Catesby's delight and amazement at the cooperation, sagacity and ludicrous industry of the beetle. His amusement is revealed in his description of their "breech foremost" progress and in his account of the beetles "and the balls . . . tumbling and rolling one over another down . . . little eminences"; his piquant fascination is reflected in his calling their "noses" excellent and in his evocation of flights of beetles descending on dung "just fallen from Man or Beast," a vivid and particular, if disgusting, detail. His amusement at the whole incredible phenomenon and the metaphorical and satiric potential of the situation is underlined by his explicit appreciation of the irony of one beetle's name: "These are commonly called *King Tumble-Turds,* tho' by what appears they assume no preeminence, but without distinction partake of the like dirty drudgery with the rest." One is reminded of Thoreau's "battle of the ants" and the mixture of fascination at particular nature and derogation of mankind contained in that piece. Catesby was even more capable than Bartram of creating "real toads" in "imaginary gardens."

Entertaining as are some of the prose selections from *Carolina,* the prevailing tone is rational, responsible, skeptical. As Carver and Bartram distrusted the French, Catesby distrusts Spanish reports of the "Politeness" and accomplishment of Mexican Indians in the "more abstruse Arts of Sculpture and Architecture" and thinks those Indians "in reality . . . only a numerous Herd of defenceless *Indians,* and still . . . perfect *Barbarians*" (An Account, vii–viii). He seems to distrust the popular belief that rattlesnakes charm their prey (2: 41), and elsewhere dismisses "as remote of Reason, as the Ethereal Region is from the Aereal," the notion that swallows hibernate under water.[6] His purpose is to educate, to inform, albeit pleasingly. Consistent with that aim, then, he questions popular fancies.

Also consistent with his aim to instruct is his inclusion of information collected and published by others. He improves the worth of *Carolina* as a reference book by numerous references to other works and Latin nomenclature supplied by Sherard and others, by quoting passages from other natural philosophers, and even, as in the essay on the "Tumble-Turd," by recalling the remarks of the ancients, Pliny and Aristotle. Part of this responsible inclusiveness is also indicated in his frequent use of non-Latin nomenclature. Peter Collinson once complained that Linnaeus's novel nomenclature perplexed "the delightful science of Botany with changing names that have been well received, and adding new names quite unknown to us." Thus, in retaining and using common names like "Pole-Cat," "Bald Eagle," "Goat-Sucker," and "Mock-Bird" as well as appending Latin names, Catesby extended the usefulness and appeal of his book.[7]

In some instances he even seems to have relished the local diction, such as "Licking hole Church" (An Account, vi), named for a natural salt lick near which it was built. One might even detect in his use of Indian matter a pride in being one of the initiated; in his account of the skunk, Catesby reports that the "*Indians . . .* esteem their Flesh a Dainty, of which I have eat, and found it well tasted" (2:62). While such lagniappes contribute to the reader's pleasure and convey the author's enjoyment of field research, they must finally be seen as intentional contributions to knowledge, just as were Carver's report of the edibility of moose-lip (T, 448) and Bartram's report that the water of a creek "smelled like bilge water" and tasted "sweetish and loathesome" (D, 40).

The Iconography of Mark Catesby

Mark Catesby's plates are certainly his most impressive contribution to nature reportage and, for the purposes of this study, his most significant contribution. In his 220 plates he limns this animal and that plant so exactly as to add graphically to the cabinets of the virtuosi; but he renders them in combinations that stand almost more as portraits than as illustrations of natural history. Illustrations such as Catesby's fulfilled for the eighteenth century the role that photography fills today—that of exact representation. Catesby's, however, go far beyond this plain and rational function. As he portrays a blue jay screaming from his perch on a Bay-leaved Smilax (fig. 3), a catbird reaching for a mosquito while perched on an *Alni folia Americana* (fig. 4), a chipmunk eating a wild nutmeg beneath a Mastic tree branch (fig. 5), or scarab beetles practicing their "oeconomical employment" beneath spotted yellow lilies (fig. 2), he creates poetic visions of the cosmos of nature; that is, to borrow Irwin Panofsky's distinctions, Catesby's plates exist both as "documents" and as "monuments," both as "references" to natural or cultural *facts* and as integrated symbolic "propositions" of *value* and *meaning*.[8]

This dual quality of *Carolina* has never been adequately studied, for scientists tend to find symbolic elements distracting or irrelevant if not destructive of accuracy, while to artists such items as the supplemental fangs and rattles of the rattlesnake plate (fig. 15) and the "lifesize" outline of the bald eagle's head (fig. 1) seem distractions. Even eighteenth-century virtuosi sometimes failed to appreciate Catesby's achievement. In his *Notes on the State of Virginia* (1785), Thomas Jefferson inserted a table of common American birds in which the parallel lists of Catesby, Linnaeus and Buffon shared equal prominence, testifying to Jefferson's dependence on *Carolina* as the only complete, reliable, *illustrated* natural history of America. At the same time however, Jefferson introduced the table with a warning that Catesby's "drawings are better as to form and attitude than coloring, which is generally too high." Thus Jefferson contributed to the already growing tendency to take a narrow view of the function of nature reportage. Dr. Alexander Garden of Charles Town similarly, if much more strongly, criticized Catesby in a letter to Linnaeus (1760). Eager to establish his own reputation abroad and never hesitant to criticize the work of others, Garden ridiculed inaccuracies in Catesby's art and charged:

It is sufficiently evident that his sole object was to make showy figures of the productions of Nature, rather than give correct and accurate representations. This is rather to invent than to describe. It is indulging the fancies of his own brain, instead of contemplating and observing the beautiful works of God.[9]

Though intending to blast Catesby's reputation, Garden articulates better even than Catesby's modern admirers or eighteenth-century supporters the special quality of invention in his work that helps to raise it above the level of mere documentation. In their criticism of *Carolina,* Jefferson and Garden foreshadowed the tendency of modern scholarship to invest its critical energies in scrutinizing *Carolina*'s representational accuracy instead of its symbolic content. Indeed, this traditional approach has become so well established that neither Catesby's biographers, Frick and Stearns, nor sympathetic art historians have been able to respond freshly to the beauty and variety of his art or to appreciate fully the cultural significance of his iconography.

The documentary aspects of Catesby's art are what have chiefly engaged the attention of critics like Elsa Guerdrum Allen and the interest of Catesby's biographers. They have carefully evaluated his usefulness as a scientific recorder and have generally reached a favorable verdict. There is no reason to duplicate their efforts here. Catesby's meticulous engraving techniques, his concern for exact color, his determination to paint from nature and his textual comments on the specimens portrayed establish representational accuracy as one of his conscious intentions—the equivalent of Carver's and Bartram's intentional verbal accuracy. All three are conscious practitioners of a style "intelligible and explicit": Catesby in line, color, space; Carver and Bartram in words, tone, format.

Useful as the comments of Allen, Frick and Stearns are to the study of Catesby's plates as documents, they contribute little to any systematic exploration of the affective content and symbolic meaning of his plates. To interpret a symbolic statement one needs some "map" of the artist's grammar, semantic field, characteristic diction and syntax. One must know something of the artist's taste for metaphor and be attuned to his individual tone, must be aware of the several ways he combines borrowed and novel elements into a finished invention, and not simply spy out "plagiarism"—a term less applicable anyway to graphic than to verbal art. In short, if one is to build a useful appreciation of Catesby's art, one must lay a careful foundation of generalizations about his affective and cognitive organization, patterns of practice, characteristic devices.

Catesby is often compared to Audubon because he set himself the task of portraying all the American birds. Elsa Guerdrum Allen's excellent contribution to Catesby scholarship treats only his "ornithological" plates; the Frick and Stearns biography of Catesby, while it treats all aspects of his document, emphasizes the ornithological by their epithet, "The Colonial Audubon." Specialization has of course increasingly dominated natural science in the last two centuries and it is inevitable that Catesby should be categorically fragmented if he is to be remembered at all in the history of science circles. But Catesby was a generalist, not a specialist, and this cannot be stressed too much. For instance, while the one hundred plates in Volume One of *Carolina* are indeed "bird plates" in that each contains a bird, they are also plates in which many botanical specimens and some nonavifaunal elements appear. Each botanical specimen is identified in the text, and the bark, leaves, fruit, flowers and roots are carefully drawn. Often, as much and occasionally more text is devoted to the plant as to the bird. There is even a double principle of order evident in certain series of plates; for example, while plates 16 through 21 feature eight species of woodpecker, plates 16 through 23 also portray eight types of oak tree, each with bark, leaves and acorns clearly figured. This is not to quibble; a basic perspective is involved. If we can look at the plates of *Carolina* freshly (ignoring or systematically discounting categorical "imperatives" imposed on *Carolina* or on "reality" by others), we will see not bird plates or fish plates but masses of plates which *include* birds, landscapes, insects, bits of domestic architecture, plants, fish, serpents, amphibians, nuts, seeds, mammals and schematic sketches of anatomical minutiae.

It has not been sufficiently appreciated that Catesby was not merely a clever artist-ornithologist who initiated the practice of adding floral backgrounds to bird images, but an artist who consistently juxtaposed images from nature—sometimes a bird and a relevant plant, sometimes a snake and a vine, even a fish and a terrestrial plant; sometimes a beetle and a lily, a sumac and a wasp, or a chipmunk and a nut. Audubon entitled his work *The Birds of America;* Catesby entitled his, *The Natural History of Carolina, Florida and the Bahama Islands: Containing the Figures of Birds, Beasts, Fishes, Serpents, Insects and Plants: Particularly the Forest-Trees, Shrubs, and other Plants, not hitherto described, or very incorrectly figured by Authors. Together with their descriptions in English and French. To which are added, Observations on the Air, Soil, and Waters: With Remarks upon Agriculture, Grain, Pulse, Roots, &c. To the whole is prefixed a new and correct Map of the Countries treated of.* The difference is not academic or merely a reflection of publishing conventions. Catesby's book is intentionally sweeping, various, plenitudinous; the work of a trained botanist who ranged over the whole of nature and delighted in reporting travel conditions, Indian customs, reptilian phenomena, agricultural potential of soils, speculations on the "universal deluge," and personal anecdotes as well as botanical and ornithological observations. In much the same spirit Bartram and Carver delighted years later in natural phenomena outside their own specialties; Bartram in the deluge, Indian customs, agricultural potential; Carver in moose-lip and sand cherries, great caves and clear water, shining mountains and Indian heroines.

Nothing short of color reproduction, in folio size, can adequately suggest the rich variety of Catesby's art. Some plates strike one as symmetrical, ordered, static (fig. 11); others as dynamic, asymmetrical and active (fig. 3). Some contain images of single specimens (fig. 15); others, of two (fig. 23); several, of three or more (fig. 4). Many are icon-like in their formal beauty, simplicity and strength (figs. 7 and 8); a few are ludicrous (fig. 9). Some, like that of the rattlesnake which includes details of fangs and rattles (fig. 15), remind one most of illustrations for the *Philosophical Transactions* of the Royal Society. Many are as decoratively charming as a medieval book illumination (fig. 13); some are "naturalistic" scenes like that of the chipmunk eating a nut (fig. 5); still others are clearly the product of an imagistic imagination rather than a naturalistic one, e.g., the flamingo pictured before a disproportionate piece of underwater coral (fig. 8). The content of the pictures is equally varied and playful. Birds stand for remarkably engaging portraits; snakes entwine themselves in writhing plants, crabs rest symmetrically on blank backgrounds like astrological icons, moths hold down the corners of flower portraits, ichneumon wasps and crickets add minute decoration; polecats, black squirrels, alligators, and scarab beetles occupy centers of interest. The point is that both in style and in content the plates of *Carolina* bespeak not a dry and consistent representationalism and settled esthetic but a restless fullness of interest and variety of invention.

One intuits that the plates are monuments in Panofsky's sense of the term—intentionally created images with symbolic presence as well as documentary content. Even so, between intuition of artistry and convincing exposition of that judgment, a critical chasm often exists. Logically, if the intuition is valid, one should be able to extract and marshall explicit visual evidence into a convincing argument. With someone like Catesby, who has been neglected by trained art critics, many such chasms exist. Frick and Stearns, for example, infer from their awareness of Catesby's financial difficulties that his decision to group plants and animals together in single plates was essentially an economic one that allowed him to pack double or triple the material into a restricted space. If economics alone motivated Catesby's composition, that would undermine the plausibility of any claim that particular combinations of flora and fauna make symbolic statements, that juxtapositions of plant and animal images are a "poetic" expression of his vision of nature. The question is, then, were the combinations of images in Catesby's plates dictated by economic, naturalistic or esthetic considerations, or by a mixture of these? Can evidence be found to indicate that esthetic considerations operated at all; for if it can, it opens the door to an analysis of Catesby's symbolic voice.[10]

In two plates particularly Catesby does combine separate elements as if to save expense only. Plate 36 (Volume One) includes a "Snow Bird" (slate-colored junco), a "Broom-rape" and a "Toad Stool." Plate 10 of the Appendix includes a life-size heron's head with five separate insects pictured around it. Neither plate has any more compositional unity than a page from a naturalist's sketchbook. The two plates are the graphic equivalents, perhaps, of the lists of specimens that occur throughout Carver's and Bartram's journals. However, the majority of the plates in *Carolina* are composed not of such random items, but of items made to belong together. Catesby's most frequent method of combining items is illustrated by the portrait of the bluebird (fig. 6). It is a balanced, unified, "framed" piece of art which in mood and composition communicates the artist's feeling for his subject as well as accurately documenting its morphological configurations. The bird stands on a pedestal, as it were—in this case the conventional, eighteenth-century, bird-artist's device of the sawn-off stump. Framing the bird and confining the viewer's eye to the inner space is a canopy of *Smilax non spinosa*, delicately rhythmic in contrast to the bird's central stasis. Visually this particular combination of bird, conventional stump and floral frame is a striking work of art; the artist has selected both conventional and novel images and synthesized them into a unified composition. The plate's simplicity, quiet mood, meticulous rendering and two-dimensional flatness makes it an eighteenth-century icon with the *Smilax* delineating a niche in which the bird, both beautiful in itself and a sign of beauty in nature, stands on its pedestal.

A very different bird portrait is that of the blue jay (fig. 3). A jay, his tail cocked, screams down from his perch of "Bay-leaved Smilax." The *Smilax* occupies the left border of the plate, and the bird's body line and attention direct the viewer's interest to the right. The plate is asymmetrical, unbalanced, active—yet the image is a finished one. The artist would seem to have adjusted his composition to the "personality" of the subject, for if ever there was a bird that evokes action and disorder, it is the sentinel jay that shrieks warnings and imprecations from its perch at intruders. Different yet, is the plate of the cardinal perched amid hickory leaves (fig. 7). The contrast of the leaves' green with the bird's red is dramatic, but the bird's central location in the plate and the bent-over tips of the surrounding hickory leaves draw the eye into the center from every border. The plate thus achieves a powerful centripetal unity of composition that contradicts or balances the tension between the opposed colors, red and green.

We have become so conditioned to Audubon's and later ornithological illustrators' conventional grouping of birds with ecologically relevant plants and insects that we are apt to overlook the philosophical and artistic importance of Catesby's invention of this practice. He is said by Mrs. Allen to have "laid the foundation of popular ornithology" by "inaugurating the combination of attractive lifelike portraits with appropriate environment, and short narrative accounts" ("American Ornithology," p. 474), and one need only compare Catesby's plates to the bird portraits in Eleazar Albin's *A Natural History of Birds* (1731–38) to appreciate the artistic effect of Catesby's invention. Albin's birds are frequently beautiful, but his backgrounds are sterile, conventional, dwarfed landscapes that contribute little to his pictures. Ultimately the most convincing evidence of Catesby's esthetic motivation as he combined elements for his plates exists in those plates which demonstrate that ecological representationalism was not inevitably a determining factor in his selection. In plate 73, Volume One, a flamingo stands upright against what appears to be a stylized tree, bare of twigs and leaves, growing out of the ground (fig. 8). Only by reading the description in the text does one discover that the *Keratophyton* of the background is an underwater growth that, in spite of its apparent size, grows only "about two feet" high. Apparently the *Keratophyton* must be interpreted not as a natural setting for the flamingo but as a compositional device that accentuates the tallness and linearity of the bird. Oddly disturbing as this combination is to the intellect, its visual impact is effective and beautiful.[11]

A plate that is even more peculiar than that of the flamingo is the final plate in *Carolina,* which pictures a buffalo dwarfed, apparently, by an acacia sprig (fig. 9). This plate disturbs the eye conditioned to Renaissance conventions of perspective. The buffalo is located near the bottom of the plate and appears to rub himself against a broken tree trunk. The size and position of each lead one to perceive that they exist in a plane some distance from the viewer. However, the one branch that grows out of the trunk and overarches the buffalo appears, from the relative size of its leaves and flowers, to occupy a plane much closer to the viewer. Catesby has violated the conventions of perspective and figured two items, presumably equidistant from the viewer, in different scales. Whatever the viewer may think of the plate, it is a grotesque failure by Renaissance standards of perspective. Clearly then, Catesby has either failed in his art, or he does not inevitably accept "natural" perspective as a basic requirement of his art. That the second conclusion is at least partially true can be supported from the evidence of other plates.

While no other plates in *Carolina* thus affront the eye conditioned by Renaissance standards, many simply ignore three-dimensionalism. They exist on a single plane and remind one of non-naturalistic book illuminations of medieval times or of twentieth-century works that reject illusions of depth as essential to art. One of Catesby's most striking fish portraits is that of the sole (fig. 10). Centered on the page and filling a space devoid of background material, it is less a representation of a live fish than it is a self-contained image existing on paper. Were it wiggling its tail or seen against a background of seaweed, it would seem more to direct the viewer's attention through its image to a reality outside the picture. As it is, subject and image are perfectly unified. One might object that a flat sole is a subject that makes it difficult to introduce elements of depth, and the objection is valid. Not all Catesby's fish plates are so successful. As documents they are occasionally flawed by the absence of a dorsal or pectoral fin; as three-dimensional illusions they are sometimes disturbing. Many show what is obviously intended as a shadow beneath them, as if resting on the ground when drawn. The picture of the "Bone Fish" is especially odd in that the fish is pictured against a background of "Sea Feather" coral as if floating in water, but carries a shadow of greenish crosshatching under it which suggests that it was lying on the mud when painted (2: 13). Thus it seems to float in watery space with a corporeal shadow attached to its belly. One can say from the evidence of the fish plates that Catesby never settled the iconographic problems of fish portraiture. He was a slave to useless shadows in many of them; he ignored depth in others.

Other of his fish plates are striking pieces of art. The portrait of the "Globe Fish," *Cornus*, and *An Phaseolus* (fig. 11) combines diverse elements, as do his bird plates, but combines them without regard for ecological naturalism and without illusions of the third dimension. The result is one of his most decorative, unified fish plates. The fish is figured at the bottom of the plate; the plants spread vertically up the center of the page but in no real sense are they perceived as behind the fish. The colors and tones of the fish and plants are identical—the dark green of the fish's back recurs in a seed pod of the plant, the lighter green of the plant's leaves reappears in the fish's coloring, and even the clean paper color that fills the spaces between the leaves of the plant appears again in the fish's light belly. Elements of color, line, figure and blank space unify the separate images as they were never unified in nature. The plate is a sort of graphic equivalent of poems in which puns and ambiguous verbal symbols add a synthetic continuity to naturally distinct elements.

If economic expediency, ecological naturalism and Renaissance conventions of perspective are not clear and consistent elements in his art, what guiding principle can be inferred? Perhaps simply to celebrate beauty and significance in nature with any and all appropriate artistic devices. Catesby uses representational accuracy, ecological combinations, and conventions of the fine arts if he wishes, but when it serves, he abandons them easily. The richness and variety this freedom allows him can be exemplified by examining several specific plates. The plates that strike the viewer as least naturalistically conceived are those of snakes. Writhing in "twining" plants, the snakes coil into perfectly concentric circles, twist in decorative but unnatural knots or cross the page in quadruple, hairpin bends. The "Corn Snake" of plate 55 (fig. 12) is viewed from above as it moves across the plate from left to right, but it is superimposed upon a representation of the *Viscum Caryophylloides* which is seen from a horizontal perspective. The exposure of the bulb and root hairs is conventional and need not indicate that the plant has been uprooted and laid on its side beneath the snake. Further evidence that artistic rather than naturalistic considerations have functioned is that the stalk of the plant is bent so that the entire plant fits within the frame of the plate.

In the portrait of the harmless "Green Snake" Catesby demonstrates a delight in the decorative possibilities of snakes that has little to do with "natural" facts (fig. 13). The snake is one of "Nature's People" for whom colonists apparently, to quote Emily Dickinson, felt "a transport of cordiality"; they often "carry them in their bosoms," we are told by Catesby. The *Cassena vera* is a plant from which the Indians make spring tonic, Catesby says. Thus the plate presents intentionally, I infer, a vision of the peaceful and decorative garden of wild nature. This snake does not threaten; it decorates the world. Representative of the less congenial aspects of snakedom, however, is a portrait of the "Black Snake" (fig. 14). Catesby tells us that the actual black snake, while truculent, is harmless to man; indeed it is positively beneficial, eating rats and even rattlesnakes. But the snake of the portrait strikes a threatening pose and in the line of its body suggests a tension that is not conveyed by the corn snake or green snake, and it evokes, in Emily Dickinson's words again, "zero at the bone."

Of all the snakes Catesby portrayed, only the rattlesnake appears without the artistic complement of a floral background or prop (fig. 15). It is coiled in the middle of the plate, its head raised for attack in one corner, its tail as a warning in the other. In the corner beneath the head is a detail of a fang; in the corner below the tail, two details of rattles, one cutaway. As a document the plate is exact, recording not only the figure of the snake but details of its anatomy that interested virtuosi. But as a monument, the portrait of the rattlesnake suggests an emotional response by the artist different from that evoked by other snakes, even other poisonous snakes. If one compares this portrait to the others of *Carolina,* the inference that the rattlesnake occupies a niche in nature distinct from other fauna, even other snakes is compelling. It is the only snake pictured without a natural context. Perhaps, then, Catesby has created an iconographic equivalent for the conceptual and emotional uniqueness of the rattlesnake. Other snakes may be capable of pleasurable, decorative integration in the design of the cosmos; the rattlesnake resists. If rattlesnakes are symbolic of that in the new world which does not fit some providential usefulness or purpose, then Catesby's portrait is the graphic expression of that uneasy truth.[12]

The rattlesnake *was* a disturbing symbol to Americans, as the frequency, variety, content and tone of their reports of it suggest. One detects in their accounts an intensity of emotion out of proportion to the actual threat from the snakes. John Bartram, mild-mannered and rational, recounts with obvious satisfaction the vivid violence of his attack on a rattlesnake. Catesby, who must have seen many rattlesnakes, accentuated the horror by reporting that one was found in his bed. Franklin chose the rattlesnake as a symbol of terrific danger in his "exporting of Felons to the Colonies." What was it about the rattlesnake in particular that excited such emotion? Not its snakeness alone nor its deadly potential. Perhaps more than these, it was the intolerable symbolic ambiguity of rattlesnakes that bothered these eighteenth-century rationalists and lovers of nature. There were other dangerous beasts in the woods, but none that so mocked man's cosmic optimism by flaunting a symbol of nature's concern for man's safety as an accompaniment to attack. Like the whiteness of the whale in *Moby-Dick,* the two ends of the snake signify opposite cosmic lessons for man, and they are joined in one snake.[13]

The majority of Catesby's plates celebrate a less horrific nature, one that is benign, decorative, colorful and useful—even amusing. They portray a cosmos of nature in which man is comfortable and can admire the usefulness, beauty and lessons of providential design. In plate 73, Volume Two, a black squirrel rests on its haunches eating a nut; by his black contrast he emphasizes the pure beauty and color of the "Yellow Lady's-Slipper" (fig. 16). In another plate (fig. 5), a chipmunk ("Ground squirrel") gnaws a wild nutmeg beneath a Mastic tree in a vignette of peaceful, self-composed, innocent industry. Several plates demonstrate a gentle sadness, a sentiment increasingly common later in the century and fully exploited by sentimental poets and engravers in the nineteenth century. The robin reposes in death, protected by a snakeroot vine (fig. 17); a dead "Yellow-rump," probably a myrtle warbler, hangs by one leg from a spider web attached to a hellebore (1: 58).[14]

A different sentiment, one of amused admiration for the variety and rational design in nature, is conveyed by the picture of the "Tumble-Turd" beetles pursuing their "oeconomical employment" on the ground beneath an umbrella of "lilium" (fig. 2). It is a fanciful combination of images of conflicting connotation balanced in one picture. The lily stalk rises straight up the middle of the page and bursts into bloom. On the dirt at the base of the stalk, two beetles work, one on each side: one trundles "breech foremost" a ball of dung; the other appears to be searching for dung. Compositionally the plate is balanced and symmetrical. The blooms direct the eye toward the beetles; the horizontal plane of the earth leads the eye from the beetles back to the lily stalk and up to the blooms. The lilies and beetles are separate, but formally yoked. The plate is hardly naturalistic: it is, instead, the graphic equivalent of a poem, a composition in which selected, separate, vividly rendered images are coupled. One thinks at first of the mixture of tension and resolution in Pope's couplets, but the meticulous particularity of the images of the beetle and lily surpass even the visual particularity of Pope's nature poetry, and the tone suggests a metaphysical quality of wit. Catesby's fancy has been engaged by the concept of yoking "Tumble-Turds" and lilies, as if to symbolize a world. I am tempted to infer that the picture is a graphic metaphor of satiric intent, the subject of which is men who ludicrously devote all their attention to "oeconomical employment" and ignore beauty in the garden of nature. There is no explicit textual support for this inference, but the frustration of naturalists with colonial indifference to natural history was widespread. At any rate, the subject matter of the plate and its composition demonstrate a wit and fancy seldom found in natural-history art.

The eighteenth century was a time of imitation, paraphrase, and translation in the fine arts. It was also an age of plagiarism and piracy. So too in natural philosophy. But while literary critics have carefully discriminated several degrees of borrowing and have distinguished between ethical and esthetic considerations, neither historians of science nor art critics have cared enough to rationalize similarly the criticism of the art and literature of natural philosophy. Nevertheless, if Catesby's (or any other nature reporter's) "art" is to be studied fully, the significance and the effect of borrowing must be considered. ("Borrowing" and "transformations" of course characterize all cultural behavior, but special codes prescribe how and when to borrow—especially in "the arts.") The task is crucial, since condensation, imitation, paraphrase and plagiarism occur widely in the literature and art of eighteenth-century science. Several of Catesby's plates provide instructive exercises in the sort of criticism that is needed generally, and are also beautiful enough to warrant the effort. Catesby was a pioneer, an innovator. He established conventions for others to follow. But like Carver and Bartram he was an inconsistent, hesitant genius. Like them, in attempting to create order out of chaos, he sometimes created combinations of conventional and novel elements that were ludicrously bad. He was working in a field as yet uncharted by convention, and sometimes lost his nerve or failed to appreciate his own genius. Two examples from *Carolina* illustrate his problem as he tried to balance the conflicting demands of convention and novelty. One is a disappointing failure; the other, a triumph of his artful combination of borrowed and original elements.

One of Catesby's most clumsy artistic failures, as embarrassingly bad as John Bartram's excursions into high style, is the headpiece he designed for *Carolina* (fig. 18). It is, however, a revealing and meaningful failure. The headpiece is clearly a conventional rococo device—a decorative combination of organic and geometric elements that intertwine to fill the space and to frame a cartouche, topped by a bust. It has been made grotesque, however, by Catesby's placement in the cartouche of a tangle of Indian implements and weapons and by the substitution of an Indian's head for the usual classical bust. The Indian seems to have an unnaturally long neck, no shoulders, and a sad expression; his hair looks like a combination of conventional portrait wig and Mohawk brush-cut into which three feathers, one rather limp, have been stuck. The implements figured include an arrow, a bow, a string of beads, a quiver, a quirt, a spear, a tomahawk, a peace pipe, a war club and one unidentified object, perhaps a sceptre. (Many of these items separately derive from John White's illustrations for de Bry. See note 15.) While the rococo elements are adequate by themselves, the implements mediocre, and the Indian's head crude, it is the whole that is intolerably jumbled and contradictory. Carver and Bartram had difficulty enough with Indians in plain prose; Catesby's attempt to fuse Indian material with highly conventional devices of book decoration is a ludicrous failure. How, one wonders, did he not dare, did he not have the self-assurance and the awareness of potential beauty to create a headpiece from *Smilax* vines, butterflies, snakes and birds of prey. His later plates show that intricate and sinuous decoration could be made from vines and that compositions of symmetry could be created from butterflies and crabs. One can only regret his failure of nerve or failure of vision, and conclude that like Carver and Bartram, he was insufficiently self-conscious of the novelty of his contribution to culture to extend his genius into the established areas of book decoration.

Catesby's plate of the "Land-Crab" is certainly one of his most striking portraits, and one of the most significant to this discussion of his art (fig. 19). It demonstrates that he could create symmetrical, decorative compositions by combining original elements with elements carefully copied from other artists. The crab is viewed from above, centered on the page, and its claws curve semicircularly upward. The design on its back contributes to the symmetrical effect. Overarching and complementing with a semicircular pattern of its own the upward sweep of the crab's claws is a "Tapia" sprig, one berry of which is lightly pinched in the smaller of the crab's two claws. The green of the leaves sets off the orange of the crab; the mass and coherence of the crab is emphasized by the fragile lines of the plant. The objective balance of the crab's symmetry is countered by the asymmetrical plant and by the location of the crab left of the center of the plate. The result is a masterpiece of simplicity, strength and exciting tension controlled by artistry.

"Masterpiece" implies artful manipulation and conscious composition by the creator, as well as successful results. The crab portrait is one of the works from *Carolina* for which known precedents exist and from which, then, some reconstruction of Catesby's imagination may be attempted. Doubtless there is a degree of inevitable symmetry in crabs' appearances. Possibly, also, conventional astrological images of crabs fed Catesby's imagination. In the case of this particular plate, however, a third and operative source of influence is known. After his return to England from America, Catesby had available to him at the library of his patron, Sir Hans Sloane, the water color sketches made by John White at Roanoke in the sixteenth century. From comparisons of Catesby's crab and White's (made possible by Paul Hulton's and David Beers Quinn's extraordinary *American Drawings of John White; 1587–1590*), it is clear that Catesby's crab is a copy of White's. Here, then, is "borrowing" as literal as any of Carver's; yet to label Catesby a plagiarist and dismiss the "Land-Crab" from consideration would be as serious an error as it was with Carver. Both men sought to add to the worth of their books by incorporating into their own work what had been done by others; both acted in a manner consistent with the ethics of their time.

Passing from ethics to esthetics, questions remain: What did the artist seek to achieve? What did the artist contribute? That Catesby sought to supplement the documentary worth of White's crab by putting it in a natural context is clear from the text and from the plate. Catesby added the "Tapia." He could have suggested a context with a bit of seaweed or dune grass. Instead, he added a specific piece of relevant flora for which there is no precedent in White's sketch. The "Tapia," Catesby says in the text, is a twenty-foot high tree upon the fruit of which the crabs like to feed. He has borrowed the crab from White and has integrated it into a new composition thoroughly original in effect. But something of Catesby's esthetic sense can also be deduced from the plate. We know from other plates in *Carolina* that a "Tapia" berry could have been shown separately in a corner like a rattlesnake fang or like the luna moth in the *Philadelphus* plate (fig. 21). Instead he artistically linked the crab and the plant by opening the crab's smaller claw so that it pinched a berry of the plant. The specimens are joined. We know also from the flamingo plate and the buffalo plate that he could have drawn a whole tree. Instead he selected a sprig. Since this decision was controlled neither by the White drawing nor by the demands of his intention to add a "Tapia," his choice of a sprig may be assumed to be esthetic. Consequently the effects of tension and decorative framing are Catesby's own. His selection of a model to copy was felicitous; his choice of context, appropriate; his composition and execution of the whole, effective.[15]

Still another instance of Catesby's borrowing is instructive. It reveals his humility, his honesty, and ultimately the particular genius of his characteristic artistic mode. Two of the plates (2: 61, 96) are signed by Georg Dionysius Ehret (1710–70), the great German engraver of Linnaeus's plates, who came to England while Catesby was working on *Carolina*. Catesby's inclusion of two plates by Ehret suggests his admiration for the man and probably his belief that Ehret's art was superior to his own. The *Magnolia amplissimo* (fig. 20) is not signed by Ehret but in "all probability," according to Frick and Stearns, is "modelled" on Ehret's plate of the same specimen. At any rate, it is a good illustration of the visual impact of the signed Ehret plates. Frick and Stearns, who consider Ehret's artistry far superior to Catesby's, praise Catesby for his willingness to learn

from Ehret the usefulness of crosshatching to indicate depth
and shading. They view his willingness to "learn" as a virtue;
one could just as well interpret this willingness as a lack of
confidence in his own genius. Similarly, their declaration of
Ehret's superiority, while justified from one perspective, betrays
their failure to sense the particular virtues of Catesby's genius.[16]

Frick and Stearns judge Ehret's *Magnolia altissima* to be
"undoubtedly the most striking picture" in *Carolina*. In the context of *Carolina*, Ehret's plates are indeed striking; their dark,
full mass arrests the eye as would a swatch of florid chintz
thrown in among samples of crewel work. Ehret's portraits fill
the entire space within the frame; the flowers and leaves are
boldly large and all blank spaces are filled in with brown cross-
hatchings. No white space remains, as in Catesby's characteristic
plates. In contrast, Catesby's flower portraits appear simple. In
his portrait of the sumac, for instance, an isolated sample of
sumac, separated from any distracting context, is set off by a
lone wasp figured in the upper-left corner (fig. 23). In his *Philadelphus* portrait (fig. 21) and *Catesbaea* portrait (fig. 22), delicate
traceries of stems, leaves and blooms appear against white backgrounds and are similarly set off by luna moths and butterflies,
respectively. Whether the Ehret-like *Magnolia amplissimo* or
the sumac portrait is "better art" is at least moot; that they betray distinctly different intentions and convey different meanings is certain. The *Magnolia amplissimo* is conceived as a win-
dow, as it were, through which one sees a portion of complex
and full reality. Implicitly, with six of its eight leaves, it leads
the viewer's mind and eye out beyond the frame of the picture
to external reality. Catesby's sumac plate is conceived differently, as a consciously arranged display of a wasp and a piece
of sumac which symbolically suggest nature but create no illusion of its actual, amazing fullness. The sumac plate is like a short
lyric poem in which meaning is suggested by formal juxtapositions and distinct images. The Ehret plates and Catesby's imitation of Ehret serve as foils for the sharp brilliance of Catesby's
characteristic plates. Not the least important result of this contrast is an increased and sharpened appreciation of the particular
spirit of Catesby's cosmos of nature, a pre-evolutionary one in
which species are the result of separate creation. For Catesby,
the separateness, variety and plenitude of nature are delightful
reflections of providential design to be celebrated in line, color,

shape. His biographers miss the point when they praise Catesby as "usually ecologically correct" and apologize for his posing a Mexican bird with a Carolina plant. They assume that his aim was, or ought to have been, a realism consistent with modern conceptions of the ecological coherence of nature. This erroneous assumption makes them blind to, or at least verbally unable to express their admiration for, his considerable symbolic accomplishments. While there may be foreshadowings of ecological coherence in Catesby's plates there is also a vivid celebration of distinction, individual virtue and symbolic meaning in the "productions of nature."

At his best, Catesby created icons—strikingly simple, removed from distracting context, meticulously rendered and delightfully illuminated and decorated. Birds of prey, traditionally symbols of the spirit, are portrayed in solitary strength; harmless snakes deny their evil reputations by entwining themselves intricately in innocent and useful plants; a rattlesnake appears ominously separate from floral mitigations of its mortal danger with its head and rattles raised; "Tumble-Turds" practice their "oeconomical employment" beneath a benign umbrella of yellow lilies; a robin reposes in peaceful death on a tree stump over which arches a protecting "Snake-Root." In his passion for viewing, passion for colors, passion for accurate delineation and passion for juxtaposing flora and fauna, Catesby created images to symbolize and celebrate the cosmos of nature he perceived as surely as did Thomas Cole, Frederick Church or Winslow Homer later. Having done so, he made a significant contribution to early American thought and art, symbolically rendering the culture's worldspace and particular integration.

A "literalist of imagination"

Nature reportage in the eighteenth century was an international endeavor; French, Spanish, German, Dutch and Swedish as well as English and American virtuosi published works of natural philosophy, illustrated and not. No assessment of Catesby's special genius or estimate of his place in the spectrum of natural-history illustrators could be complete without some comparison of his work to others'. Many such works exist; some are modest, some ambitious, and some splendid beyond imagining unless one has seen them in the rare book rooms and special collections of fine libraries and museums. Probably many of these could throw into relief the special virtues of Catesby's art, but the two I have found most relevant and suggestive are the four-volume *Locupletissimi rerum naturalium* . . . (Amsterdam, 1734–65) of Albertus Seba and the smaller but engaging *A Natural History of Birds* (London, 1731–38) by Eleazar Albin.

Albertus Seba, according to the *Enciclopedia Universal Illustrada Europeo-Americana* (Madrid, 1927), was a Dutch merchant, trained in pharmacy, who travelled widely for the Indian Company and after having acquired a considerable fortune, employed part of it to form a collection *"de las producciones más raras de la naturaleza."* Much of his collection of natural rarities is reproduced in *Locupletissimi* and is published in Latin, Dutch and French, with illustrations by Pieter Tanje, a commercial book illustrator. The four-volume work staggers the viewer with its completeness, variety and sumptuous beauty. "Precious and costly" Catesby's *Carolina* may have been, but compared to Seba's morocco-bound folio volumes, it is modest and restricted in subject matter. The contrast is not quite as extreme as between Cook's *Voyages* and Carver's *Travels*, but it is considerable. Seba's specimens were collected from the whole world and to turn his pages is to be confronted with hundreds of snakes at a glance, many on a page, each brilliantly tinted, precisely engraved and writhing in decorative if unnatural poses. One meets also pictures of two-headed fetuses and rarities similar to those one finds in the *Philosophical Transactions* of the Royal Society. By contrast, Catesby's plates of select productions of nature, his twenty snake portraits, his half-dozen mammals, seem a modest addition to the virtuosi's book shelves. Even so, once the awe at the richness of Seba's collection is past, Catesby's plates are more artistically satisfying. Whereas Tanje's illustrations of Seba's collection of snakes remind one of the color plates in unabridged dictionaries and encyclopedias, crammed full of brilliant examples, each of Catesby's snakes occupies an individual plate. Consequently, Catesby's portraits seem to recognize an individual existence and personality in his subjects while Tanje's strike one as awesome representations of natural specimens, but not as portraits. Catesby allows one to contemplate a "Coach-Whip Snake" by itself and to learn from the facing text that the Indians report that these can "by a Jerk of their Tail, separate a Man in two

Parts" (2: 54). Seba's work may be more inclusive, even more valuable as a herpetological reference work, but Catesby's is a superior artistic accomplishment, a far superior symbolic entree to the culture that is its context and to the personal negotiation of the nature/culture dialectic by one nature reporter. Catesby's portrait of the rattlesnake is as fine a document as any virtuoso could desire, but it does more than inform the viewer of the appearance of the snake: it conveys the ambiguous and fascinating symbolic quality of the snake as well.[17]

Catesby, like Thoreau and Bartram and Carver, was a thoroughly interested and engaged reporter, not a disinterested scientific collector. His value for students of American culture lies not in the completeness of *Carolina*, which is incomplete by modern or even by Wilson's or Audubon's standards, but in its particular attention to individual specimens, to their emotional impact as well as their visual appearance. Seba's work is primarily a graphic representation of one peripatetic virtuoso's collection of rarities; Catesby's, the monument of a sensitive and rational man's passionate view of nature in the new world. Seba's work presents regiments of specimens; Catesby's, studies of memorable types.

Eleazar Albin (fl. 1713–59) is also an ideal subject for comparison to Catesby. An early water colorist and devoted, if superficial, naturalist, his *Natural History of Birds* appeared in the same year as Catesby's first installment of *Carolina*. In 1719 he and Catesby had both been considered by James Brydges, Duke of Chadros, for the naturalist's post in an expedition he was mounting to Africa, but Albin turned it down, and Catesby, after wavering, went instead to America. Albin's interest in nature extended, as did Catesby's, beyond birds to insects, spiders, fish; his *Natural History of Spiders* (1736) reveals his interest in the relationship between birds and their eight-legged parasites. Above all he was author and engraver of several remarkably attractive volumes on birds which, he announced on their title page, had been "curiously engraven from the life" by him and "carefully colour'd by his daughter . . . and self." He usually achieved likenesses equal to Catesby's (sometimes better) but tinted his engravings more opaquely so that the color obscured the engraved lines beneath. Consequently his plates look more like paintings than engravings, and while just as meticulous a rendering of patterns results, Catesby's plates show a minute documentary accuracy missing in Albin's. Nevertheless, both men were devoted to the careful and accurate transcription of particular beauty in the birds they drew.[18]

Where Catesby and Albin differ most is in their treatment of backgrounds and props for their prime subjects. Catesby usually chose specific plants for backgrounds and perches. Albin preferred only generalized clumps of turf, sawn-off stumps or dwarfed landscapes and sailing ships conventional then to such art. His owls invariably stand on stumps; his ducks and geese, on pieces of turf surrounded by water and fringed with stylized, miniature cattails. Catesby also perched owls on stumps and drew tiny ships to accompany his aquatic birds, but while Albin never rose above this "shorthand," Catesby usually did. In the majority of cases Catesby created precise, particular, carefully delineated and recognizable suggestions of environment for his portraits, even to including a cutaway chimney for his "American Swallow" (chimney swift) and a section of thatched shed roof for his purple martin (App. 8; 1: 51).

One gathers from viewing the Albin plates that one bird at a time challenged his skill. His aim was to celebrate and record the colors and shape of the bird by itself; its characteristic surroundings or activity were beyond his detached interest as an artist. Catesby, on the other hand, intrigues the viewer by combining flora and fauna. One is convinced that the bird actively exists in a real place—the backgrounds may not always be ecologically correct, but they are artistically alive and accurate. Albin's specimens generally repose as if mounted in a virtuoso's cabinet; Catesby's more often are posed as they might be in life: his kingfisher swallows a fish (1: 69); his catbird threatens an insect (fig. 4); his nuthatches climb head-first down a trunk (1: 22). One of the most animated plates in *Carolina* features an eagle diving for a fish, the fish itself tumbling upside down, and another eagle gliding on outstretched wings in the background (fig. 1). Compositionally, Catesby's plates may often exhibit a balanced symmetry, but his birds frequently appear in characteristic, active poses against particular and accurate backgrounds. His birds are more "real toads" than Albin's; his backgrounds are imaginary gardens of selected but thoroughly convincing images. Albin's imaginary gardens are as trite as the scenes on theater fire curtains or the landscape backdrops in a photographer's studio.

Catesby more than Tanje or Albin is a "literalist of imagination." He is meticulously literal in his engraving techniques, draftsmanship, tints, poses and backgrounds. This is not to say that his colors, draftsmanship and backgrounds cannot be challenged by modern naturalists; they can and have been. The point is that he intended to be literal in his depiction of form, color, pose and background detail—literal, as a poet might be, so as to convey with concrete images the texture, particular beauty and emotional tone of the subject. That he is a literalist of imagination and creates imaginary gardens with his art is especially apparent from comparisons of his plates with Tanje's and Albin's. By adding a real lily to the portrait of the "Tumble-Turds," he implies a world instead of a virtuoso's cabinet; by adding a *Cassena vera* to the green snake plate, he implies a cosmos.

His value for students of Anglo-American culture, then, lies both in the fullness of *Carolina* and in the integrity of its individual pictures, the best of which are not realistic portraits of nature but icons—delightfully illuminated and decorated, yet strikingly simple; meticulously rendered, yet free of distracting elements. Singly the plates illustrate the particular beauty, texture and meaning of nature; as a group they symbolize and celebrate the intense and discontinuous variety of the cosmos of nature Catesby perceived. That special eighteenth-century cosmos of nature exists no more, but the persistence of nature and man's continuing delight in it allows Catesby to communicate across the gap of two centuries.

But more than that, Catesby created connections between culture and nature which enrich our own present. His works are capable of intruding their incarnate being, their "affecting presence," on our consciousness in a world far removed in time and space from his own. They intrude not merely as mediating structures (as vehicles transporting exotic meaning from then to now), but as immediating constructs. Which is to say that they appresent themselves to us as live presences, face-to-face, as it were, and that Catesby has in effect, then, augmented the store of being in the world today, though he himself is long gone; and that is worth making the most of.[19]

Illustrations

1
Bald Eagle. *Aquila Capite Albo.*
[1:1] Note: Numbers in brackets refer to the illustration's location,
   by volume and page, in *Carolina.*

2
Tumble-Turds. *Scarobæus Pilularis Americanus.*
Lilium sive Martagon Canadense, *floribus magis flavis, non reflexis.*
[App.:11]

3
Blue Jay. *Pica glandaria cærulea cristata.*
Bay-leaved Smilax. *Smilax lævis Lauri folio baccis nigris.*
[1:15]

Muscicapa vertice nigro. *Alni-folia &c.*

4
Cat-Bird. *Muscicapa Vertice Nigro.*
Alni folia Americana serrata, floribus pentapetalis albis, in
  spicam dispositis. Pluk. Phyt. Tab. 115, f. 1.
[1:66]

5
Ground Squirrel. *Sciurus Striatus.*
Mastic Tree. *Cornus, foliis Laurinis, fructu majore luteo.*
[2:75]

6
Blue Bird. *Rubicula Americana Cærulea.*
Smilax non spinosa, humilis, folio Aristolochiæ, baccis rubris.
[1:47]

7
Red Bird. *Coccothraustes Rubra.*
Hiccory Tree. *Nux Juglans alba Virginiensis.* Park Theat. 1414.
Pignut. *Nux Juglans Carolinensis fructu minimo putamine levi.*
[1:38]

8
Flamingo. *Phoenicopterus Bahamensis.*
Keratophyton Dichotomum fuscum.
[1:73]

9
Bison. *Bison Americanus.*
Acacia. *Pseudo Acacia hispida floribus roseis.*
[App.:20]

Solea lunata &c

10
Sole. *Solea Lunata et Punctata.*
[2:27]

Cornus &c

An Phaseolus &c   Orbis &c.

11
Globe Fish. *Orbis Lævis Variegatus.*
Cornus, foliis Salicis Laureæ acuminatis; floribus albis; fructu
  Sassafras.
An Phaseolus minor lactescens flore purpureo. Hist. Jam.
  Vol. I.—162.
[2:28]

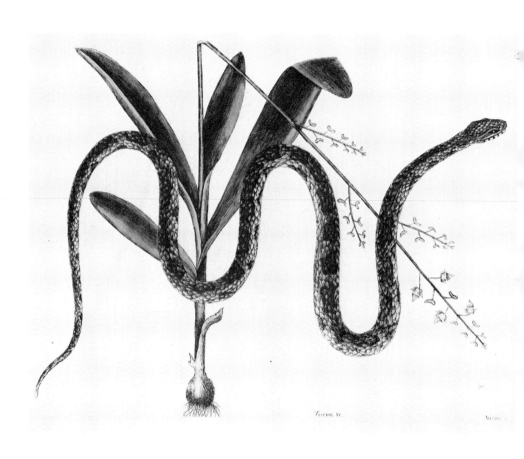

12
Corn Snake. *Anguis e rubro & albo varius.*
Viscum Caryophylloides ramosum, floribus minimis albis.
[2:55]

13
Green Snake. *Anguis Viridis.*
Cassena vera Floridanorum, Arbuscula baccifera Alaterni facie,
  foliis alternatim sitis, tetrapyrene. Pluk. Mant.
[2:57]

14
Black Snake. *Anguis Niger.*
Frutex Rubo similis, non spinosus, capreolatus; fructu racemoso
  cæruleo Mori-formi.
[2:48]

The Section
of a Rattle.

A Rattle of
Twenty four
joynts.

A Tooth.

Vipera caudisona.

15
Rattle-Snake. *Vipera Caudisona Americana.*
[2:41]

16
Black Squirrel. *Sciurus Niger*.
Yellow Lady's Slipper. *Calceolus* Marianus *glaber, Petalis angustis*.
   Pet. H. 1. 5. Raii, Hist. III. App. 243. *Vid*. Pluk. Tab. 418. fig. 2.
[2:73]

17
Fieldfare of Carolina. *Turdus Pilaris, Migratorius.*
Snake-Root of Virginia. *Aristolochia pistolochia seu Serpentaria*
Virginiana caule nodoso. Pluk. Alma. p. 50. Tab. 148.
[1:29]

18
Headpiece with cartouche from title page of *Carolina*.

Tapia &c.

Cancer terrestris.

19
Land Crab. *Cancer terrestris, Cuniculos sub terra agens.* Nat. Hist.
  Jam. Vol. I. T. 11.
Tapia trifolia fructu majore oblongo.
[2:32]

20
Umbrella Tree. *Magnolia, amplissimo flore albo, fructu coccineo.*
[2:80]

84

21
Philadelphus flore albo majore indoro.
Smilax non Spinosa baccis rubris.
Four Eye'd Night Butterfly. *Phalæna plumata caudata,* Caro-
liniana, *virescens oculata.* Pet. Mus. p. 69. No. 733.
[2:84]

22

Catesbæa. Lycium Catesbeii, Authore D. Gronovio. *Frutex Spino-*
sus Buxi foliis, plurimis simul nascentibus; flore tetrapetaloide,
pendulo, sordide flavo, tubo longissimo; fructu ovali croceo,
semina parva continente.
Papilio caudatus Carolinianus; fuscus, striis pallescentibus; linea
& maculis sanguineis subtus ornatus. Pet. Mus. p. 50. No. 508.
[2:100]

23
Ichneumon Fly. *Vespa Ichneumon Tripilis, Pensyvaniensis.*
Sumac. *Rhus glabrum Panicula speciosa coccinea.*
[App.:4]

Tumble-Turds and Lilies

Beneath the yellow lilies, Tumble-Turds
Pursue their oeconomical employment.
Two scarabae with thorax shields of crimson
And green gold, with ribbed and shining
Deep green wings and black horns recurved
Backward, search the dirt beneath the blooms
For dung just fallen hot from man or beast.
And from this excrement, one beetle makes
A ball to house an egg and rolls it backward
Off to its interment under lilies.

Up from the middle of the dirt a lily
Thrusts its stalk and spreads its yellow blooms
In floral canopy as if to make
A niche for these most humble artisans,
As if asserting general harmony
Between such opposite particulars.

East of Eden even beetles work
In dust and plant their generations in
The filth to be reborn through warm decay
To green gold irridescence. The Garden's lost,
But beauty, purpose, even dignity
Persist where minute order seems design
And filth is spangled with such brilliant shards.

David Scofield Wilson, 1961

Epilogue

An epilogue is a chance to have an additional say: a say *upon,* a say *over* and *above,* a say *around,* and a say *toward.* In this case it is a chance to expand briefly on matters of nature and culture largely outside the special purview of this study and to comment on matters of culture study, *per se,* the dimensions and tendencies of which are implicit, perhaps, but not always spelled out in the body of the work. It is a chance to make some extra connections and to address readers concerned about the drift of modern American (and world) culture and anxious somehow to make sense of the world and in it.

Today we live in a world, a socially constructed "modern cosmos," in which nature and culture often seem at odds. The cosmos the nature reporters helped knit together in the eighteenth century has come unravelled. Two centuries of scientific and industrial growth have broken nature into categorical bits and pieces and into commodities to be bought and sold. And two centuries of getting a living and getting along in the face of developing modern conditions have broken down, or at least made increasingly implausible, the kind of satisfactory union of nature and culture embraced by the nature reporters in their work and enacted more articulately by people like Franklin, Jefferson, Benjamin Rush; more critically by those like Thoreau; more artistically by Whitman or Emerson or Frank Lloyd Wright. The list is incomplete, even haphazard, but serves to suggest how problematic a holistic vision and unified way of living have become over the intervening centuries.

On a global scale, nature seems out to get us, to pay us back for DDT, aerosol cans, over-population, single cropping, over-exploitation of fossil fuels, and massive pollution of the sea and air and land. Certified scientists and free-lance Jeremiahs join in "lining out" scenarios of ecological doom. At home, extravagant energy is put into keeping nature in its place, in lawns, in parks, out of the house, out of sight, out of touch, out of smell: weeding, air conditioning, pest controlling, deodorizing. Or into straightening teeth, regulating conception, improving hybrids, schooling children. As a matter of course, whatever is born seems to be taken as something to be fixed up rather than celebrated for what it is.

Our world is so massively different in organization and texture from the world the nature reporters inhabited and helped to make that comparisons, especially normative ones, are in most respects exercises in futility. We cannot go back even if we would, and I would not. The past is passed and cannot "teach us" in any obvious way what we ought to do today. Why then poke around in and try to make sense of a piece of the past? Because the past is what we make of it, and we make up ourselves in the act of giving it meaning. If we make it mean *nothing*, we incorporate the zero into our own being. To make it mean *something* it did not before we started alters the present. It makes plausible certain human ways of being in the world and makes more real the recognition that the most taken-for-granted ingredients of everyday life and the most self-evident truths of our symbolic universe were not always true or obvious. By seeing that past reality was once constructed and has since been dismantled, we are better able to grasp that our own real and urgent world is simultaneously a matter of fact and a matter of consent: a dialectical dynamic in which *what is* massively affects what can or might be, and in which what we take or refuse to take as true affects finally what is. Which is to say, something *can* be done.

It is hard to say what should be done in response to apocalyptic scenarios. Either they are true predictions—in which case only a massive, worldwide, economic, technological and cultural revolution promises any effective remedy—or they are fantasies and rhetoric, culture-specific strategies of symbolic universe maintenance. It is clear enough from a culture-studies point of view that at some level all held beliefs are *ethnotruths*, beliefs that grow out of specific material and spiritual conditions and contribute, dialectically, both to the continuing plausibility of the taken-for-granted cosmos and its processes, and to the implausibility of radically alternate ways of making sense out of possibilities apparent in the present. I do not mean by this relativizing to discount the dread embodied in the prophecies of disaster. I mean instead to illuminate the ways in which the sense of being trapped resembles the sense of impending doom antecedent to major paradigmatic revolutions, whether religious, scientific, social, or cultural. The perception that "the centre cannot hold" (Yeats, "The Second Coming") is often right, in one obvious way: not as a prediction of the future, perhaps, but as a heightened awareness of disintegration experienced here and now. The future aside, owning up to that domestic and personal disorder is a necessary if not sufficient first step on the road to revolution or redemption. The words sound melodramatic but after all, at the root level, mean simply "turning around" or "getting back." Thus, I take heart in a perverse way from the apocalyptic pronouncements; they show that others share my sense that there are better ways to act in the world than are supported by habit and by the values and meanings incarnate in taken-for-granted institutions and roles in America.

Getting down to the more mundane, everyday realities, down to the existential matters of living out lives of quiet desperation in the world or doing something about it, I take heart from the lives of the nature reporters. They witness the possibility of breaking out of a received world and making their way in a new manner, practically and enjoyably. Not without upset, but without turning renegade or going crazy in the end. They trusted their own uneasiness, and trusted themselves and other like-minded activists to be capable of discovering the truth and realizing it in useful work. They were at once both responsive and responsible to the world.

If I look at today's nature/culture complex expectantly and search for those who might qualify as descendants of the nature reporters, I find certain self-reliant activists who, *mutatis mutandis*, seem to embody a similar responsiveness and responsibility, practicality and liveliness. On the one hand, the loose collectivity of "natural-living" reporters exhibits a similar eagerness to make *ad hoc* sense out of received technologies and to make new ways of cashing in on the underlying processes of the world. They devise ways to use solar energy, wind power, natural predation chains, sky lights, organic fertilizers, and the like. And they test their schemes on themselves and neighbors and publish their findings, their "know-how" and their projections in newsletters as well as in books, catalogues and periodicals. They suffer similar ridicule and well-meant "therapy" at the hands of established professional, political and commercial custodians of traditional culture. And they suffer a similar distrust by ordinary citizens.

Quite different at first glance, on the other hand, are the "human-potential" reporters; but the advocates of T'ai Chi, meditation, zen awareness, gestalt therapy, Rolfing, est, T.A., and too many more to list, perceive themselves as working to restore some balance between inborn and acquired needs, between nature and nurture, in short, at the very core of being.

The human-potential and natural-living reporters, like the nature reporters before them, demonstrate that neither the objective nor subjective "patterns-in-experience" must remain fixed. Reality may indeed be socially constructed, "all set up," but it can also be reconstructed, day by day and bit by bit. That may not yet add up to revolution or redemption on a global scale, or even on a culture-wide scale, but it often amounts to such on a personal level; and the collective action of these reporters and others is beginning to build role-specific knowledge and institutional arrangements which will one day make alternative ways of negotiating nature and culture more real, and surely more lasting.

And finally—the past and the prototypical reporters aside —the affecting presence of works by Catesby or Bartram or White or Dudley (or many others) intrude in my world, here and now. They hold power I was unaware of before I encountered them. I trust they do for others, too. And their power empowers me. I am encouraged as an artist and a writer by their presence, just as I am by my friends and co-workers. And as the works present tumble-turds, yellowish wasps, snapping turtles, groundsell down, and the like, I find my world enriched, not only by these static presences but by the spiders and jays and trees I now notice more intensely than I ever did before. I am more alertly alive as a result, and I am grateful for that.

It should be evident from the foregoing that at the very least American Culture Studies is to me an interdisciplinary endeavor which calls into play the methods of iconographer, literary critic, historian, biographer, poet, geographer, anthropologist, sociologist, or even natural scientist if the occasion warrants. This has been the way of American Studies over the decades since its beginnings in academic teaching and research programs in the 1930s and 1940s. Those of us who now identify ourselves as doing American Culture Studies mean by that extended denomination to place the culture concept at the center of our theory and methodology as an organizing principle.

In the context of the present work I have taken American Culture Studies as a discipline to my researches and my articulation of findings, which means: I sought to encompass both large, systematic connections as well as affective and cognitive connections made by individuals under the stress of needing to make sense *of* and *in* a received world which did not quite fit them without strain. I have taken the past as present in the design of ongoing interrelationships as well as present in the more immediate sense of Armstrong's phrase, "the affecting presence." Furthermore, the culture concept, and especially the principle of the dialectical relationship between objective reality and subjective consciousness, helps make sense of the way a particular historical nexus of things and ideas impels and makes plausible certain roles, truths, values, and meanings which individual actors manifest, and of how those same actors have the power to transform to a greater or lesser degree the very reality they inhabit. Whatever is socially constructed may be socially reconstructed. To me American Culture Studies is logically an inclusive rather than exclusive discipline. To others it might appear to be logically indistinguishable from humanistic anthropology or sociology.

Some of the differences that exist in fact derive vertically from the history of American Studies and other academic programs, and horizontally out of the matter-of-fact organization, funding and peopling of academic departments and all that goes along with such professionalism. The more important differences, however, grow out of the logic of a situation in which Americans engage in American Culture Studies.

At the very heart of the matter is *interestedness,* the acknowledgment that I cannot refuse for long to look upon matters of importance as if I had no stake in them. Or as if my own cognitive and affective organization were not, in part at least, wrapped up in the dynamics of my object of study. The *reductio ad absurdum* of the aim to engage objectively and disinterestedly in reflexive culture study is the picture of a man studying his image in the mirror but taking it as an artifact only and none of himself; especially since everyone else can see plainly what he is about and how he is fooling himself, to speak kindly, or exhibiting bad faith, to speak more censoriously. Better to own up at once and lay out as clearly as possible as much as one understands of his own (in this case my) principles of selection, judgment, management of data, and all the rest. This makes for what some perceive as a mixed tone, or an odd socio-humanistic mix of voices, in works of American Culture Studies; it is no accident, but a matter of good principle; even good sense. Consequently, reflexive American Culture Studies requires a vigorous and continuous sensitivity to one's own processes as well as to the data at hand from out there in the world. This I have tried to display in my writing, sometimes in extended footnotes to the text, sometimes in the text itself, and of course in this epilogue and "A Personal Narrative" that follows.

Another characteristic of American Culture Studies grows out of proper interestedness, on the one hand, and out of the decision to attend especially to how culture works, on the other. This means the search for topics that reveal not the dull inability of subjects caught by conditions, but for topics that display the way Americans negotiate, transcend, or actively give in to conditions in which they have a stake and which they contribute to prolonging or ending by their acts. This point is more difficult to articulate satisfactorily in the abstract. What it comes down to, perhaps, is the insistence on doing research that *empowers* rather than demoralizes oneself and others inhabiting the culture. And that means taking a stand on political and ethical questions, at some level, whether abstractly and implicitly only, or concretely and avowedly.

This concerned stance has implications for how and where to search for topics. Suppose at the outset that the primary confidence trick perpetrated upon a population in the name of culture is the one in which takens are made to seem givens. Called reification, alienation, legitimation, theodicy, rhetoric, therapy, or any of a number of other names depending on exactly what is being "put over," it comes down to the matter of making a particular *ethnoworld* appear to be a "seamless web" or a "cosmic egg." In the service then of debunking such fictions made to seem real, more often than not American Culture Studies searches for "cracks in the cosmic egg," to use Joseph Chilton Pearce's happy metaphor, and then attends especially to how those cracks are opened up to the advantage of being, or mended in the interests of keeping the world closed. Topics like this are abundant since "the biologically intrinsic world-openness of human existence," in the words of Berger and Luckmann, is of necessity in constant tension with the "relative world-closedness" of the social order as given, and the active tendency of it to close down and close in.

The nature reporters slipped through a crack and refused to be shut out, and that is part of what attracts me to them. And their work had fallen through a crack in the professionally maintained archives of American culture, and that attracted me too. Which accounts in part for the sometimes gleeful, and I hope not self-righteous, explicit attention to what I took to be discipline-induced oversights in some of those who cared about and worked on these matters before I got to them. No one has total immunity to culture-induced shortsightedness or idiosyncratic folly, and I want, at last, to acknowledge those many who have supported me, helped me see and think better, and helped me see when I was off and appreciate where I was on.

Several individuals stand out as having singly influenced my work by encouraging me at what I take to have been crucial spots in its development: Brom Weber for first encouraging the interest I took in these materials; Mary C. Turpie for shepherding me through the agony of producing a dissertation from them; Robert Merideth whose remarkable mix of ruthless criticism and friendly support and advice drove me substantially to reconceptualize much of what I had done earlier; and Bonnie Wilson, who knew enough to hold back when premature criticism would have thrown me into despair or provoked stubborn and stupid resistance, but who always voiced her doubts eventually and persevered in criticizing ugly language or obscure argument; who resisted well but also offered solutions and made valuable contributions; and who shares my interest in nature and culture.

In a less direct way I have been encouraged by three friends and their work: David Robertson (University of California, Davis) for his daring to make photographs of natural objects speak of "black holes" and their meaning and feeling; by Paul David Johnson (Wabash College) for sharing my interests in nature writing and nature; and by Loren Owings (UCD Library) for his witness that bibliographic scholarship and making sense of one's own caring for the land go together.

I want also to recognize students too numerous to mention for their research assistance, and other students who contributed by reading and criticizing portions of the manuscript and by listening and responding in class, by resisting or assenting to what I claimed about nature reporters and their reportage. I reached a wider audience with the publication of some of the Catesby material in *Eighteenth-Century Studies,* and thank especially Robert Hopkins (editor) and Charlyn Fishman for easing that transition from manuscript to print. Similarly, Adam Horvath, Leo Marx, Douglas Vaughan, Malcolm Call and (especially) Richard Martin for their responsive reading at one stage or another of composition and for alert and thoughtful, and often right, editorial suggestions. And finally, I want to thank those friends or colleagues I induced at one time or another to read and comment on chapter drafts or the whole: James Woodress, Marvin Fisher, Janice Zita Grover, Suzanne Mikesell, and especially Leo Lemay; and thank those at the University of Minnesota who so long ago served on my Ph.D. committtee: John Howe, Edward Griffin, Bernard Bowron, John Parker, and Mary Turpie, director.

I am conscious of a debt to certain institutions *qua* institutions: to the Division of Archives and Manuscripts, Minnesota Historical Society, for permission to quote from their typescripts of the British Museum Carver manuscript material; to the Minneapolis Athenaeum for permission to photograph their copy of Catesby's *Carolina* for the frontispiece; to the Bancroft Library, University of California at Berkeley, for the photographs reproduced here as black-and-white illustrations; and to the Regents of the University of California for continued research grants, a Summer Faculty Fellowship (1970) and a special subvention that made the color frontispiece a reality.

While institutions are by nature impersonal, one of the

pleasures of doing research in them is discovering individuals there who contribute imaginative, expert advice to floundering patrons. To the Minnesota Historical Society and its staff I am indebted for much I have learned about Jonathan Carver. I owe a particular debt of gratitude to the staff of the James Ford Bell Collection at the University of Minnesota: to Carol Urness who went out of her way to introduce me to the treasures of that collection; and to John Parker, Curator, who read my work on Carver and let me read his. I am particularly grateful as well to Betty L. Engebretson, Minneapolis Athenaeum Librarian, for her imaginative, interested and expert assistance with the Catesby and Seba materials. In a more general way, I acknowledge the help of the staffs at the libraries of the University of California, Davis, the University of Minnesota, the Henry E. Huntington Library; but particularly to the librarians of the Biological-Medical Library, the Art Library, and the Natural History Museum at the University of Minnesota whose knowledge of special bibliography and particular books saved me hours of time.

A Personal Narrative

I have often envied naturalists and wished I were one. As a child I took photographs of goldfinches nesting and great blue herons fishing, watched eclipses, took plaster casts of deer and wildcat tracks. Of course others already knew everything I found out and had put it into books or museums tens or hundreds of years before. There was no call for my fresh (to me) intelligence of nature's diversity and harmonies. My father loved to paint water colors of nature, and my mother, to walk through it. My grandfather Wilson had been a fisherman and mushroom collector, birdwatcher and stargazer, and when his eyes failed, he became a geology buff and maker of sundials. My grandfather Scofield, a retired electrical engineer, had turned to genealogy, mathematical puzzles, inventing and woodworking by the time I knew him. Each cared about the workings of nature in his or her own way, but nobody needed their fund of experience either. I suppose that this book is in part a tribute to them, in part a personal working out and justification of my caring about nature and about people who cared about it.

I first paid attention to historical science when I worked as a page and shelver in the Biological-Medical Library of the University of Minnesota. Growing up with an artist led me to take the illustrations in old herbals and anatomies for curiously delightful pieces of art rather than "visual aids" to science. In time I came to value almost in direct proportion as they were "contaminated science" these curiosities which had somehow slipped through the cracks of normal disciplinary attention.

Later I found a way to slip between the cracks, myself. I learned of the program in American Studies. The people in American Studies at Minnesota not only allowed my interest in flawed science, but actually encouraged and even legitimated my attention to it. I wrote papers on Gunn's Medicine, on colonial science, and on Carver's *Travels*. The program offered more

than simple latitude, of course: names to call groups of things—
genres, styles, movements, myths. The Brooklyn Bridge, the
Frontier, Andrew Jackson all fit in.

When it came time to get serious and write a dissertation,
I decided to rescue Carver and others from anomalous status
and make a niche for them in American culture. I set out to be
interdisciplinary, to join the methods of the iconographer and
the literary critic to materials of historical science. I found my
title in metaphor: "The Streaks of the Tulip" (1968). My
choice of subtitle pretty well indicates how I set out to place and
legitimate what I studied: "The Literary Aspects of Eighteenth-
Century American Natural Philosophy."

After coming to the University of California, Davis, and
falling into a split appointment (American Studies and English),
I let "The Streaks" sit a couple of years before setting out once
more, this time to make a proper book of it. My first task was to
find a better name for the people I wrote about, a better name
for what they did. Basic naming work. "Nature reporter" and
"nature reportage" solved that problem. The more important
one, the sort of split between English and American Studies, was
more difficult to solve. Reading Thomas Kuhn, Leslie White,
Peter Berger, and Thomas Luckmann, among others, in connection with teaching American Studies classes, taught me perspectives I had not had earlier and gave me concepts and language
with which to clarify and articulate cultural processes. In the
meantime I continued to make the link between science and
early American literature by collaborating with Karl Keller on
a special issue of *Early American Literature,* "Science and Literature Issue" (Winter 1973), with the encouragement of
Everett Emerson. I kept reading, teaching, talking and corresponding: reading anthropology, theology, history, psychology;
teaching courses on religion in American culture and qualita-
tive methods of culture studies, on the one hand, and courses on
colonial American literature and "The Nature Writer," on the
other; getting help and criticism from readers of chapters I was
working to revise. I wanted discipline and direction.

Working through questions of discipline and culture
studies with Jay Mechling and Robert Merideth for an essay on
that subject ("American Culture Studies: The Discipline and
the Curriculum," *American Quarterly,* October 1973) got me off

dead center, started me down the path of reconceptualizing and reformulating the larger design of what I had begun so long ago. I moved to a full appointment in American Studies and took American Culture Studies as discipline. The rest has been working out exactly what that discipline means, in concrete terms, for the patterns and particulars that are the subject for this book, and for the enterprise of American Culture Studies generally. I prefer *discipline* as a verb. "Getting it all together" is an act, after all, and only a thing after the fact. I feel liberated rather than tyrannized by my struggle for discipline and know from experience that others may too.

Abbreviations

AHAL	Moses Coit Tyler, *A History of American Literature: 1607–1765* (1878; rpt. New York, 1962).
APST	American Philosophical Society *Transactions.*
D	John Bartram, "Diary of a Journey Through the Carolinas, Georgia, and Florida from July 1, 1765, to April 10, 1766," APST, n.s. 33, Pt. 1 (1942), pp. 1–122.
CHAL	*Cambridge History of American Literature* (New York, 1923).
DAB	*Dictionary of American Biography.*
DNB	*Dictionary of National Biography.*
J1	Jonathan Carver, "Journals of the Travels of Jonathan Carver in the year 1766 and 1767" (British Museum Add. Mss. 8949, 8950), typed transcript at Minnesota Historical Society, St. Paul, Minn., first version.
J2	Carver, "Journals," second version.
JS	Carver, "Journals," survey log.
LHAR	Moses Coit Tyler, *The Literary History of the American Revolution, 1763–1783* (1887; rpt. New York, 1941).
LHUS	*Literary History of the United States,* ed. Spiller et al. (New York, 1953).
M	William Darlington, *Memorials of John Bartram and Humphry Marshall* (1849; rpt. New York, 1967).
O	John Bartram, *Observations on the Inhabitants, Climate, Soil, Rivers, Productions, Animals, and other matters worthy of Notice . . .* (London, 1751).
RSTP	Royal Society of London *Philosophical Transactions.*
T	Jonathan Carver, *Travels Through the Interior Parts of North America, in the Years 1766, 1767, and 1768* (London, 1781), facsimile by Ross & Haines (Minneapolis, Minn., 1956).

Notes

Chapter I

1. I am indebted to Robert Plant Armstrong for his phrase, "the affecting presence," from *The Affecting Presence: An Essay in Humanistic Anthropology* (Urbana, Ill., 1971); and also in the second paragraph (and throughout) for his description of culture as "patterns-in-experience," from *Wellspring* (Berkeley, 1975). These two books should be read by all students of American culture interested in presence both in art and in other phenomena, even though the primary subject of these books is the affecting works of Yoruba and Jogjakarta. The special worth of these books lies in the clear and compelling synthesis Armstrong effects between esthetic vitality and anthropological structures.

2. I have developed some of the present-day implications of how we pay attention to nature and culture in "Nature as Education," in *Culture as Education,* ed. Vincent Crockenberg and Richard LaBrecque (Dubuque, Ia., 1977).

3. Panofsky's study of Renaissance iconography produces a vocabulary of criticism that is useful for the articulation of the esthetic dimensions and affective meaning of documents not generally assumed to be art. See "The History of Art as a Humanistic Discipline," in *Meaning in the Visual Arts* (Garden City, N.Y., 1955), pp. 1–25. Also "Iconography and Iconology: An Introduction to the Study of Renaissance Art," ibid., pp. 26–54. Culture change and maintenance form the subjects of two theoretical books invaluable to the student of historical science and American culture: Thomas S. Kuhn, *The Structure of Scientific Revolutions* (Chicago, 1962) and Peter L. Berger and Thomas Luckmann, *The Social Construction of Reality* (Garden City, N.Y., 1967). Their influence on my work is most visible when I adopt their concepts (paradigm from Kuhn; cosmos maintenance from Berger and Luckmann), but it impregnates the whole, especially the concepts, structure and methodology of the introductory chapters.

4. Several standard treatments of the data of early exploration and discovery exist: Edward Heawood's *History of Geographical Discovery in the Seventeenth and Eighteenth Centuries* (Cambridge, England, 1912),

called "definitive" by Boise Penrose; Boise Penrose, *Travel and Discovery in the Renaissance: 1420–1620* (Cambridge, Mass., 1963); J. H. Parry, *The Age of Reconnaissance* (London, 1963); Percy G. Adams, *Travelers and Travel Liars, 1660–1800* (Berkeley, 1962).

5. No one to my knowledge has employed the terms *nature reporter* and *nature reportage* before. Generally current are *scientist, naturalist, natural historian, natural philosopher,* and *virtuoso* for the persons; *science, natural philosophy,* or *natural history* for the activity. Some seeking more descriptive precision have coined *exploring naturalist, classifying naturalist* (William Martin Smallwood, *Natural History and the American Mind* [New York, 1941]), *Linnaean traveller, field naturalist, literary observer,* and *nature writer* (Robert W. Bradford, unpublished dissertation, "Journey into Nature: American Nature Writing, 1733–1860," Syracuse University, 1957). But none of these adequately conveys the nature reporter's whole cultural function nor frees the reportage from the tyranny of conventional literary or scientific assessments of worth.

6. *The Battle of the Books* appears in *Gulliver's Travels and Other Writings,* ed. Louis A. Landa (Boston, 1960), pp. 355–80. In *Science and Imagination* (Ithaca, N.Y., 1956), Marjorie Nicolson shows how close to the mark Swift's satire of the Royal Society is in *Gulliver's Travels,* Book III. Additionally, Swift's choice of a spider to represent "the moderns" was genius: not only was the Royal Society actually fascinated by spiders about the time Swift wrote (David S. Wilson, "The Flying Spider," *Journal of the History of Ideas* 32 [1971]: 447–58), but the spider and the bee became rather complex symbols of freedom and civilization in poetry and fiction (Tony Tanner, "Notes for a Comparison between American and European Romanticism," *Journal of American Studies* 2 [1968]: 83–103). "Sweetness and light" lived on in disputes between science and the humanities (see Matthew Arnold's "Sweetness and Light," in *Culture and Anarchy* [London, 1869]). *Boswell's Life of Johnson* (London, 1960), p. 267, and *Rambler* 82 and 83 contain Johnson's disparaging remarks.

7. Reynolds' "Seventh Discourse" and Johnson's *Rasselas* appear in *Eighteenth Century Poetry and Prose,* ed. Louis I. Bredvold et al. (New York, 1956), pp. 1174–89, 707.

8. Alexander Pope, line 135 from *Essay on Criticism* (1711), widely anthologized.

Chapter II

1. This is not the place to untangle and catalogue all the strands of influence which snake in and out of my formulation of the relationship

between culture and nature, the dynamics of cultural evolution, the drama of personality and culture interaction. Alert readers will infer many debts without my help, even those of which I am largely unconscious, but I do wish to acknowledge a few works that repeatedly intruded upon my thoughts as I struggled to articulate what seems to me a plausible and "interesting" model. With regard to the contest between the conditioned and the unconditioned and the distinguishable "projects" arising out of that dialectic: René Dubos, *So Human an Animal* (New York, 1968); Peter Berger, *A Rumor of Angels* (Garden City, N.Y., 1969), and *The Sacred Canopy* (Garden City, N.Y., 1967); Harvey Cox, *The Secular City* (New York, 1965); Mircea Eliade, *The Sacred and the Profane* (New York, 1961); Edward T. Hall, *The Silent Language* (Garden City, N.Y., 1959), and *The Hidden Dimension* (Garden City, N.Y., 1966); Thomas Luckmann, *The Invisible Religion* (New York, 1967); William F. Lynch, *Christ and Prometheus* (Notre Dame, Ind., 1970); Bronislaw Malinowski, *Magic, Science and Religion* (Garden City, N.Y., 1948); Robert Merideth, "The Value of Meaningless Questions in the Study of (American) Culture," a paper delivered at the American Studies Association convention (Washington, D.C., 1971), and "A Statue in the Park," the introduction to a book in progress, "Culture Against Nature: Paul Goodman and Adam"; Murray G. Murphey, "On the Relation between Science and Religion," *American Quarterly* 20 (1968) 275–95; Richard E. Sykes, "American Studies and the Concept of Culture: A Theory and Method," *AQ* 15 (1963) 253–70; Teilhard de Chardin, *The Phenomenon of Man* (New York, 1959); Leslie A. White, *The Science of Culture* (New York, 1949). With regard to coping with and within culture, see Noam Chomsky, *Language and Mind* (New York, 1968); Erik Erikson, *Young Man Luther* (New York, 1958); J. G. A. Pocock, *Politics, Language and Time: Essays on Political Thought and History* (New York, 1971); Michael Polanyi, *Knowing and Being* (Chicago, 1969); Edward Sapir, *Culture, Language and Personality* (Berkeley, 1949); Anthony F. C. Wallace, *Culture and Personality* (New York, 1970); Benjamin Lee Whorf, *Language, Thought, and Reality* (New York, 1956).

2. The two best secondary sources of information about those who pursued natural knowledge in early America are: Raymond Phineas Stearns, *Science in the British Colonies of America* (Urbana, Ill., 1970), and Brooke Hindle, *The Pursuit of Science in Revolutionary America, 1735–1789* (Chapel Hill, 1956). Smallwood's earlier and narrower history still contains much that is valuable, especially his treatment of the passing of the "naturalist" and his long bibliography of primary and secondary sources (*Natural History,* pp. 355–424). Smallwood, a zoologist, pays scant attention, however, to the literary or artistic elements in the material he assembles (see remarks on Catesby, pp. 28–30, and

Bartram, pp. 32–35). Philip M. Hicks's well-known *Development of the Natural History Essay in American Literature* (Philadelphia, 1924) is little help since his operative criteria are narrowly literary, causing him to select Crèvecoeur as the father of the natural history essay rather than any of several earlier candidates—Jonathan Edwards, Paul Dudley, John Bartram, for instance.

3. Roy Harvey Pearce, *Colonial American Writing* (New York, 1950), pp. 16–17; brackets added by Pearce.

4. Perry Miller and Thomas H. Johnson, eds., *The Puritans,* rev. ed. (New York, 1963), pp. 122–25.

5. *New English Canaan* (1637; rpt. Washington, D.C., 1838), p. 53.

6. In *Lawson's History of North Carolina* (rpt. Richmond, Va., 1951), p. 119.

7. Ibid., xx; Colden, *History* (Ithaca, N.Y., 1969), xi–xii. Regarding style, little has been published that bears directly upon nature reportage; the prefaces and introductions to the primary sources are the best source of explicit principle while the texts themselves reveal implicit, and sometimes contradictory, standards. Several general discussions of style provide a useful background: Thomas Sprat's preference for "the language of Artizans, Countrymen, and Merchants before that of Wits or Scholars" and for a "close, naked natural way of speaking" (1668) is often taken incorrectly to be definitive of the Royal Society style, *Cambridge History of American Literature* 8: 368–69. Other CHAL essays of general utility are: A. A. Tilley, "The Essay and the Beginning of Modern English Prose," F. A. Kirkpatrick, "The Literature of Travel, 1700–1900," and W. W. Rouse Ball, "The Literature of Science." Carson S. Duncan, *The New Science and English Literature* (Menasha, Wisc., 1918), and Lawrence A. Sasek, *The Literary Temper of the English Puritans* (Baton Rouge, 1961), develop categories and vocabulary helpful to one concerned with style.

8. *Lawson's History,* p. 124.

9. With five John Winthrops prominent in American history, I embrace gratefully Stearns's practice of designating John Winthrop (1714–79) "Professor Winthrop." The many other reporters introduced briefly in this chapter are commonly included in histories of science and literature, thus relieving me of the tedious necessity to footnote each appearance of them or their works in my text. Stearns, Hindle, *Literary History of the United States* and *Dictionary of American Biography* should be consulted if additional data is desired.

10. Borland, *Homeland: A Report from the Country* (New York, 1964), p. 7.

11. Anthropologists use terms like *ethnobotany* and *ethnozoology* to describe the "science" practiced in primitive or exotic cultures. The terms have not caught on among historians of American science even though they perceive that the nature studies done a century or two

ago were "contaminated" by culture. But *ethnoscience* is the best term for what naturalists did, and even a term with heuristic virtues for the study of modern science—consider the difference between the acupunctural and analgesic approaches to pain. The illustration is instructive. For years the efficacy of acupuncture was deemed implausible in the West and taken as an instance of China's cultural lag. Now it has been rationalized in neurological terms and is scientifically "interesting" for the first time. Not a "culture lag" but a "plausibility gap" separated Western and Eastern medical scientists. A similar gap separates us from early American science. Their notion that God or Nature had so designed the world that antiveninous plants flourished precisely in the neighborhood of poisonous snakes seems implausible today, even to most nonscientists. One would think they could have seen the evidence to the contrary. But it was not a matter of perception, not a matter of failed genius or dedication, but simply that their culture apportioned differently what sort of propositions seemed interesting and plausible.

12. *Andover Review* 13 (1890): 1–19. One or another of the two versions composed by Edwards are collected in various anthologies. For a thorough discussion of Edwards's composition and its relation to the history of science and literature, see my essay cited in note 6, chap. 1.

13. Franklin's letter (25 August 1755) appears in *Benjamin Franklin, Representative Selections,* ed. Chester E. Jorgenson and Frank Luther Mott, rev. ed. (New York, 1962), pp. 272–74.

14. Smallwood, especially, devotes considerable space to museums, academies and popular science. See also Max Meisel's *A Bibliography of American Natural History* (New York, 1967), vol. 2, a Hafner reprint of the 1924 edition.

15. Crèvecoeur, "Letter XI," in *Letters From an American Farmer* (Garden City, N.Y., n.d.); Ernest Earnest, *John and William Bartram* (Philadelphia, 1940), pp. 14–15; William Darlington, *Memorials of John Bartram and Humphry Marshall* (New York, 1967; facsimile of the Philadelphia, 1849 edition), pp. 324–25.

16. RSPT 6 (1671): 2170–74; Stearns, *British Colonies,* pp. 687–90, 703–7.

17. Stearns lists colonial Fellows of the Royal Society, pp. 708–11; discusses the American Society, pp. 662, 670–2.

18. Joseph Kastner adopts Collinson's phrase as the title for his recent, welcome book, *A Species of Eternity* (New York, 1977), an engaging, finely illustrated, nonacademic account of the fraternity of naturalists (from John Ray to Asa Gray) who won their "immortality" by their service to natural history. Kastner catches the human side of their correspondence, travels, sacrifices, and triumphs and never patronizes his subjects or their science.

19. One epicycle on the model of culture advanced earlier: any culture exhibits (1) tacit and explicit formal propositions about being (ontology),

knowing (epistemology), and the end of life (eschatology); (2) customary patterns of social behavior (ethics, politics, child rearing, communication, etc.); and (3) purposeful inventions that facilitate human designs (axes, by-laws, chalices, canals, etc.). Graphically speaking these might be denominated *ontosphere, ethosphere* and *mĕkhosphere*—I coined *mĕkhosphere* [fr. *mēkhos; The American Heritage Dictionary of the English Language* s.v. "machine"] to improve for my purposes E. T. Hall's formulation of the "major triad" (*The Silent Language*), an idea manifest in varying forms in the work of many anthropologists. Interconnected, these three spheres constitute culture; separately, Faith, Society and Science, loosely speaking. Artificial, cutomary or ritual connections between one level and another characterize and cement any particular culture. The constellation formed of (1) propositions of causality, (2) patterns of cooperation between certified scientists and (3) their use of apparatus to experiment with nature distinguishes, for example, the scientists of modern, complex literate cultures from their colleagues in sacred cultures who as inspired or lone adepts divine the location of caribou with a burnt and cracked caribou scapula. See Omar Khayyam Moore, "Divination—A New Perspective," *Environment and Cultural Behavior,* ed. Andrew P. Vayda (Garden City, N.Y., 1969), pp. 121–29. Any culture must generate, maintain and may alter these interface connections, and this is part of what the new naturalists did.

About the APST; it is available in several forms; in my experience, the American Antiquarian Society's microfiche reproduction (Evans #11959) is superior to the University of Michigan's American Culture Series microfilm, some pages of which are illegible. The second volume of the *Transactions* did not appear until 1786. For parenthetical citations, the abbreviation APST is used. Whitfield J. Bell, Jr., *Early American Science, Needs and Opportunities for Study* (Williamsburg, Va., 1955), remains a provocative guide to other sources and topics worthy of study by students of colonial American culture.

20. APST 1: 117–97. R. P. McCormick, "The Royal Society, the Grape, and New Jersey," New Jersey Historical Society *Proceedings* 81 (1963): 75–83; Hindle, *Pursuit of Science,* p. 198. In addition to the lack of any information in the DAB and DNB, a search of the *International Index, Poole's Index, Nineteenth Century Reader's Guide* and *Biography Index* reveals no mention of Antill; even the article on the Antills in the NJHS *Proc.,* 3d series, 2 (1897): 25–56, fails to mention Edward Antill's contribution to APST.

21. APST 1: 1–3; *Ency. Amer.* 21 (1960): 10; Hindle, *Pursuit of Science,* pp. 170–71.

22. Winthrop, RSPT 57 (1767): 132–54. *Sibley's Harvard Graduates* 9 (1956), contains a good biography of Winthrop. Another rich source of nature reportage is *Gentleman's Magazine,* but unfortunately au-

thors of reports in that periodical are often inadequately identified; e.g., the piece on rattlesnakes *"communicated by Mr* Peter Collinson, *from a Letter of a Correspondent at Philadelphia"* and signed "J. B." Stylistic evidence leads me to accept Klauber's judgment (p. 1268. See note 13, chap. 5, below) that "J. B." is "probably" Joseph Breintnall and not Bartram, but in lieu of conclusive documentary evidence this ascription remains unproven.

Though only thirteen colonists between 1712 and 1773 were awarded the "highest scientific honor" of fellowship in the Royal Society, their membership is no index of American participation. Fellows consistently "communicated" to the Society the discoveries of others, in abstracts of letters, paraphrases and sketches of specimens. To discover all reports, one must look not only in the index for names of Americans, but also at the entries by Collinson, Franklin, and others. Failing to do this, Earnest was unable to find John Bartram's best piece (the "yellowish wasp"), which is indexed under Collinson's name rather than Bartram's. The variety of American contributions is immense. They include the arcane and the typical, the grand and the minute: the weather, the Indian, and rattlesnakes (including cures for their bite); the cicada, the moose, and darkness at Detroit; the refinement of "sand iron" into metal and maple sap into sugar; even "elephant teeth" from Ohio and "a cluster of small teeth observed at the root of each fang of a rattle snake": See RSPT: John Winthrop, 4 (1764): 185; Benjamin Franklin, 55 (1765): 182; Sir William Johnson, 63 (1773): 142; Benjamin Gale, 60 (1770): 244; Paul Dudley, 30 (1721): 292; Collinson and Bartram, 54 (1764): 65; Paul Dudley, 31 (1721): 165; James Stirling, 53 (1763): 63; Jared Eliot, 53 (1763): 56; Paul Dudley, 31 (1721): 27; George Croghan, 57 (1767): 464; and John Bartram, 41 (1739–41): 358.

23. Miller and Johnson, *The Puritans,* pp. 747–50, include Dudley on maple sugar production and lining bees. Dudley's pieces appeared in the RSPT 31 (1720–21): 27–28, 145–46, 148–50, 165–68; 32 (1722–23): 69–72, 231–32, 292–95; 33 (1724–25): 129–32, 194–200, 256–69; 34 (1726–27): 261–62; 39 (1735–36): 63–73. Miller and Johnson and the RSPT index attribute the piece on bees to Paul; Stearns does not. Authorship remains to be established, though who wrote it does not affect my argument. Those requiring an answer must weigh Stearns's logical but circumstantial argument (*British Colonies,* note 149, p. 459) against the common attribution, including the heading, "By the same Mr. Dudley," *same* meaning Paul. Rigorous and systematic stylistic analysis might help, though my own more casual stylistic comparisons support the plausibility of the traditional attribution.

24. Whether or not Melville, who read many old whaling documents, knew Dudley's work I have not discovered (I find no mention of him in Howard P. Vincent's *The Trying-Out of Moby-Dick,* Boston, 1949).

Nevertheless, some of Melville's descriptions of "killers" show notable similarities to Dudley's: (1) "He sometimes takes the great Folio whales by the lip, and hangs there like a leech, till the mighty brute is worried to death," "Cetology . . . Book II (Octavo), Chapter IV. (Killer)," *Moby-Dick,* ed. Charles Fiedelson (New York, 1964), p. 192; (2) "They bait the monster, as dogs a bull. The killers seizing the right whale by his immense, sulky lower lip, and the thrashers fastening on to his back, and beating him with sinewy tails," *Mardi,* Standard Edition (New York, 1963), vol. 1, p. 49. Still the only description I have seen that matches Dudley's for horror is Captain Charles M. Scammon's account, quoted by Joseph Wood Krutch in *The Forgotten Peninsula* (New York, 1961), pp. 176–77: "Suddenly, the high dorsal fins of a pack of Killers appeared, cutting the water like great black knives as the beasts dashed in. Utterly disregarding our ship, the Killers made straight for the Gray Whale. The beast, twice the size of the Killers, seemed paralyzed with fright. Instead of trying to get away, it turned belly up, flippers outspread, awaiting its fate. A Killer came up at full speed, forced its head into the whale's mouth and ripped out great hunks of the soft, spongy tongue. Other Killers were tearing at the throat and belly while the poor creature rolled in agony." Dudley's early accounts may have influenced Melville or Scammon, but I have no evidence beyond the similarity of imagery and treatment. More likely all three report events from nature which change little from sea to sea or century to century, and since all three share a stock of ideas, values, and images from American culture, similar happenings triggered somewhat similar descriptions.

25. The special "Science and Literature Issue" of *Early American Literature,* ed. Karl Keller and David S. Wilson (Winter, 1973), collects a number of interesting essays, several pertinent to questions addressed in this chapter: Joan Hoff Wilson, "Dancing Dogs of the Colonial Period: Women Scientists," corrects systematic and accidental oversights (of women) by historians of colonial culture and provides suggestive bibliographical leads which ought to be followed up; Lawrence Lan Sluder, "God in the Background: Edward Taylor as Naturalist," supplies additional evidence of the culturally "charged" nature of fossil investigations; the rest treat various facets of the culture/nature interchange arranged under three headings: "The Web of Culture: Intellectual, Social, Technical," "Spinning the Web: The Logical, Pathetic, and Ethical Witness," and "Maintaining the Web: or, Coping with Crisis."

Chapter III

1. An indication of the prestige of Carver's *Travels* throughout the nineteenth century is that toward the end of the century (1897), Moses Coit

Tyler devoted more than half of his chapter on traveller-explorers to Carver in *The Literary History of the American Revolution, 1763–1783*, calling Carver "an American Englishman in whom shone some of the best traits of Elizabethan Englishmen two centuries before," and praising him as an "obscure provincial captain" who "anticipated by forty years the American statesmanship which...sent Merriwether [sic] Lewis and William Clark to...the Rocky Mountains": "As a contribution to the history of inland discovery upon this continent, and especially to our materials for true and precise information concerning the 'manners, customs, religion, and language of the Indians,' Carver's book of 'Travels' is of unsurpassed value. Besides its worth for instruction, is its worth for delight; we have no other 'Indian book' more captivating than this" (LHAR 1: 150).

The editions of Carver's *Travels* (or *Three Years Travels Through the Interior Parts of North-America for More than Five Thousand Miles* as it is often titled) are best displayed in John T. Lee's "A Bibliography of Carver's Travels" and "Jonathan Carver: Additional Data," published in the *Proceedings* of the Wisconsin Historical Society (1909), pp. 143–83, and (1912), pp. 87–123, respectively; supplemented by Russell W. Fridley's "The Writings of Jonathan Carver," *Minnesota History* 34 (1954): 154–59.

No important changes occur in the body of the *Travels* between the first and third editions, and since the third is best known and includes Lettsom's misleading but significant biography of Carver, I have used it instead of the 1778 edition. Carver published a pamphlet (*A Treatise on the Culture of the Tobacco Plant...*) in 1779; in some of the later editions of the *Travels* the color plate of a tobacco plant from the *Treatise* is included in the *Travels*; e.g., the 1781 *Travels* in the James Ford Bell room collection of the University of Minnesota library. Theodore C. Blegen, *Minnesota: A History of the State* (Minneapolis, 1963), p. 69, says Carver's Falls of St. Anthony print is the first ever published.

2. *Travels*, pp. 470, 471, 475, 467, 457–64, 448, 449, 450, 479–85, 517–18, 523, 514, 503, 496.

3. Catesby, RSPT 44 (1747): 435–44; Williamson, Rittenhouse, and Bard, APST 1 (1769–71).

4. See Franklin's account of "the grand Leap of the Whale...up the Fall of Niagara," Jorgenson and Mott, *Benjamin Franklin*, pp. 315–17; Audubon's impositions upon the credulity of Rafinesque (drawing him imaginary fish) and his exploitation of the naïveté of a distinguished Edinburgh audience of naturalists (telling them tall tales about the copulatory habits of rattlesnakes: Alexander B. Adams, *John James Audubon, A Biography* [New York, 1966], pp. 173–74, 333–36).

5. The right of nature reporters to "lie" is by no means accepted by all readers. It is resisted especially by those who wish to exploit the re-

portage, looking through it to the commodities and patterns of nature
rather than at it for the intricate webs of self and cosmos that define
the unique author and his negotiation with his cultural and material
environment. The resistance has two dimensions, one professional and
one personal. Scientists and historians commonly seek "objective truth"
of some sort, and toward that end treat documents as more or less contaminated vehicles of fact; their methodologies become strategies for
decontamination or immunization, and much energy is expended ex-
posing myths or lies, detecting internal contradictions, flanking the
subjectivity of authors or themselves. Professionally, the mendacity of
reporters becomes a challenge with which to cope; personally it is frus-
trating. Tall tales, hoaxes, playful misdirections (even artifices of style
and form) become distractions, at best, affronts, at worst, for such art
empowers the writer and controls the reader, forcing him to consent
at least temporarily to the writer's fictions rather than his own. It forces
him to play by rules of tone and form he believes perverse when he
would rather the author had subscribed to rules that facilitate scholars'
work, had delivered into his hands clean fragments for interpretation
and organization according to current models of reality.

Carver was chided for his fictions and inaccuracies by many trappers,
explorers and geographers in the late-eighteenth and early-nineteenth
centuries. Each felt bound both to demonstrate his acquaintance with
Carver's earlier account and to attack it as less accurate than his own.
Though the motivation of such criticism may be transparent and its
practice fairly automatic, it is essentially constructive, building on the
past while keeping it alive.

6. Moses Coit Tyler had richly praised Carver in 1897 (see note 1 above).
Within a decade Bourne addressed the American Historical Society of
which Tyler had been a founder and denounced Carver's *Travels*, cast
doubt on Carver's literacy and accuracy, and by the use of parallel
columns demonstrated that Carver, or someone, had "plagiarized"
freely from Charlevoix and others: Bourne's paper, "The Authenticity
of Carver's 'Travels,'" was read before the Wisconsin State Historical
Society (December 29, 1904) and reported in the Society's *Bulletin of
Information*, no. 24 (Madison, 1905); it appeared later in the *Am.
Hist. Rev.* (1906), pp. 287–302. Bourne clearly delighted in exposés,
but while his excesses were personal, the task was social, (1) professionally an act integral to the logic of the professionalization of history,
(2) generally a part of the spirit of the times among artists, writers and
intellectuals bent on demythologizing American culture.

Bourne's attack ultimately "provoked," as John Parker said at the
90th annual meeting of the Historical Society of Michigan (1964), "a
series of inquiries more valuable than his own" (*The Great Lakes and
the Great Rivers: Jonathan Carver's Dream of Empire* [Lansing, 1965],
p. 1). A number of able scholars and dedicated laymen began assem-

bling from archives, libraries and museums the minutiae of genealogy, military pay records and forgotten memoirs that modern historians required to refute or mitigate the numerous charges against Carver and his book. Many of the issues disputed by these men in the first quarter-century following Bourne's attack are no longer of interest to the nonspecialist. The doubts about his early life caused by John Coakley Lettsom's misleading and inaccurate but well-intentioned "Some Account of Captain J. Carver," which he inserted into the third and posthumous edition of the *Travels* (1781), have been adequately resolved by a small but convincing bibliography of monographs and biographical essays: see especially Louise Phelps Kellogg's digest of scholarship and reappraisal, affirmative, of Carver in "The Mission of Jonathan Carver," *Wisconsin Magazine of History* 12 (1928): 127–45; William Browning's impressive assembly of genealogical and biographical data in "Early History of Jonathan Carver," *Wisconsin Magazine of History* 3 (1920): 291–305; three articles on Carver, Rogers, and Tute on the northwest passage by Thompson C. Elliott, "The Strange Case of Jonathan Carver and the Name Oregon," "The Origin of the Name Oregon," and "Jonathan Carver's Source for the Name Oregon," *Quarterly* of the Oregon Historical Society 21 (1920): 341–68; 22 (1921): 91–115; and 23 (1922): 53–67, respectively; and Milo M. Quaife's excellent if tendentious reconstruction of Carver biography and discussion of the Carver Grant and Carver's character in "Jonathan Carver and the Carver Grant," *Mississippi Valley Historical Review* 7 (1920): 3–25. The plagiarism issue raised by Bourne no longer much scandalizes Carver scholars. Blegen's interpretation of the evidence and evaluation of the charge is representative: "The second part of his book—really a separate work—drew much upon Charlevoix and other writers without acknowledgement, but modern critics sometimes forget that borrowings from other writers were common in the travel literature of the eighteenth century. The custom of the time . . . was to quote without quotation marks. Carver was one of many who indulged in the practice" (*Minnesota,* pp. 69–70).

Even more may be said. Bourne's parallel columns of quotations are damning out of context. In context, they impress one more as extracts or condensations by Carver to which he has appended or into which he has inserted his own observations. Throughout the section on the Indians (the portion most attacked), Carver cites or alludes specifically to the authors from whom he is accused of plagiarizing. If this is plagiarism, it is of a very public kind. The second part of the *Travels* is indeed a "separate work," as Blegen says. Blegen would forget it and look to the "Journal" portion, for which manuscript exists. A less surgical solution is simply to adopt a fresh perspective, view the book as one of a type that almost inevitably mixed original observation and common knowledge. This convention is adhered to even by scrupulously honest modern travellers such as Joseph Wood Krutch and Edwin Way Teale, both of whom habitually digest and incorporate the observations of others into their natural-history travels. Carver's Indian section ought to be viewed as a digest of all that he knows about the Indians, spiced by his own observations and experiences, illustrated by uncredited quotations "without quotation marks."

7. The confusion caused by "Carver's Grant," by which the Sioux were supposed to have given Carver title to a large section of eastern Minnesota and western Wisconsin, has now been resolved without prejudice to Carver's honesty. In Blegen's judgment, the grant was "patently forged" and a fraud which cannot justly be laid to Carver's own machinations; "no mention of any such grant" is made "in his book or diaries," and Carver, himself, "advanced no claims in relation to it" (*Minnesota*, pp. 69–70). But the "Grant" was part of the book, the part written by Lettsom, and readers took it seriously. President Monroe, the courts and Congress rejected the claims of his heirs to the land, but that did not kill the Grant's hold on the imagination of succeeding Americans. See Quaife, note 6, above.

8. Dr. John Coakley Lettsom's "Some Account of Captain J. Carver" is little help and a positive trap (T, 1–22). For reliable data one must turn elsewhere, but Lettsom's piece is not without interest, for it casts Carver as cultural hero, an image that generated further myths and counter-myths in succeeding eras. The DAB piece by Kellogg is good, but old, and in effect emphasizes doubts about Carver's *Travels*. The best assemblage of facts about Carver's history and the facts of his travels is certainly John Parker's *The Journals of Jonathan Carver and Related Documents, 1766–1770* (St. Paul, 1976) published by the Minnesota Historical Society Press. It also makes the substance of Carver's journals widely available for the first time.

Less essential but interesting are Victor C. Jacobsen's "John Coakley Lettsom and His Relations with Jonathan Carver," *Annals of Medical History*, n.s. 2 (1930): 208–16, and G. Hubert Smith's "Carver's Old Fortifications," *Minnesota History* 16 (1935): 152–65. Jacobsen reveals nothing new and is uncritical in his acceptance of Lettsom's biography of Carver, but what he says of Lettsom and of the Rev. Samuel Peters, friend of Carver, is entertaining. Smith compiles what has been said about Carver's "fortification" and evaluates it all thoughtfully.

Of the pre-Tyler scholarship, John Goadby Gregory's "Jonathan Carver: His Travels in the Northwest in 1766–8" in *Parkman Club Publications*, no. 5 (1896) deserves reading as one of the few items of nineteenth-century comment that approaches serious, thoughtful appraisal of Carver's worth for "delight and instruction." Other nineteenth-century articles are mostly so uncritical or outdated by twentieth-century discoveries about Carver as to be of little use.

9. Browning, "Jonathan Carver," pp. 300–301; Lee, "A Bibliography," pp. 143, 148; Lee, "Additional Data," pp. 90–94, 96; Robert C. Davis, *Encyclopedia Americana* 5 (1960): 690.

10. Lee, "Additional Data," pp. 96, 99, 102. A stunningly persuasive demonstration of this sort of homologue is presented by Sacvan Bercovitch in his piece, "Cotton Mather," in Everett Emerson's collection of essays, *Major Writers of Early American Literature* (The University of Wisconsin Press, 1972).

11. Carver, "Journal of the Travels of Jonathan Carver in the year 1766 and 1767," second version, typescript of British Museum Add. Mss. 8949, 8950 photoduplicated copies at the Division of Archives and Manuscripts, Minnesota Historical Society, St. Paul, Minnesota. Parker fashioned one narrative from these two versions and modernized orthography and other mechanics for *The Journals* (1976), making it necessary for me to continue to rely on these older and less accessible texts which make comparison between versions possible. Otherwise, Parker's edition, with its excellent notes and coherent record, is the superior source. Parenthetical notes identify first version as J1, second as J2; pagination follows MHS typescript.

12. *Travels* gives less explicit credit to Rogers, who as Governor of Michillimackinac bore responsibility for the expedition, than supporters of Rogers would like. They mistake Carver's first-person tone for arrogant scene-stealing and seem especially to resent Carver's greater fame in the nineteenth century. Rogers indeed deserves more attention than he has even now received, for he too was a rich and dynamic American type, but as a writer he never succeeded popularly as Carver did. Part of the reason is surely his ambiguous fame: part American hero, part Tory and certainly court-martialled. Kellogg's article (see note 6, above) is useful on Carver's relationship to Rogers and others (Arthur Dobbs of North Carolina) who sought the Northwest Passage; she concludes that Carver "unwittingly filched" Rogers's fame. Part of the problem is the stretch of years between 1768 and 1778. Carver praised and fully credited Rogers in his manuscripts even though Rogers was under a cloud, having been arrested on charges of treasonable relations with the French. Self-interest would have suggested less honesty. By the time Carver's *Travels* came to press, however, long after Rogers's disgrace and return to favor, Carver's reputation (even survival) was at stake— not Rogers's. It may be that Carver "unwittingly filched" Rogers's dream and fame, but Carver could not have known that his own book would become instantly popular and that Rogers's earlier published *Concise Account* and *Ponteach* would fail ever to become equally popular. He cannot really be blamed for achieving popularity and might easily be conceived to have thought Rogers's own fame well established while his own was still unrealized.

13. In addition to Kellogg, and of course Carver's *Travels,* I have profited from Parker's concise account of the Rogers/Carver/Goddard/Tute project (1965). Elliott's "Jonathan Carver's Source" publishes a "Copy of Major Rogers's Commission to Mr. Jonathan Carver Micha 12 August 1766" directing Carver to look for the "North West Passage . . . or . . . the great River Ourigan that falls into the Pacifick about Latitude Fifty" (see note 6, above).

14. Lee, "A Bibliography," pp. 152–53, 144; Parker, *The Great Lakes,* p. 9. I profited as well from a paper not available outside the Minnesota Historical Society: Oliver Wendell Holmes's unpublished summer (1924) term paper for Professor Buck, University of Minnesota: "Jonathan Carver's Original Journals," pp. 12–14. (Holmes became Executive Director of Natural History Publications Commission at the National Archives, Washington, D.C., in 1961.)

15. Parker, *The Great Lakes,* pp. 13–14; Victor C. Jacobsen, "Lettsom," pp. 208–16; Lettsom in Carver's *Travels* (T, 18–20); and Carver himself, p. 541, supply the information on his life in London. See LHAR and Lee, "A Bibliography," p. 45, for Carver's part in *The New Universal Traveller* (London, 1779).

16. Such borrowing was common. My purpose is not to "expose" Jefferson, Colden or others but to put Carver's actions in perspective; consequently I have not documented Jefferson's borrowings, or Colden's, though it can be done. For an example of how Carver borrowed from Rogers see Allan Nevins in Robert Rogers's *Ponteach: or the Savages of America* (Chicago, 1914). Compare Carver's accounts of the beaver, bear, and porcupine (T, 282, 274, 279) to Roger's accounts *(Concise Account,* 253, 259, 263).

17. The artfully natural narration of events, with anecdotes, information and speculation, characterizes countless nineteenth- and twentieth-century works as well as Carver's. Authors of literary travels as well as narratives of exploration and scientific expeditions also utilized this formal device of the eighteenth-century nature reporter. Especially apparent in narratives of expeditions like those written by Keating and others in the first half of the nineteenth century, the form persists into the twentieth century and is visible in the works of such men as Joseph Wood Krutch, Edwin Way Teale, John Steinbeck and others. John Muir followed the same plan when he larded his accounts of Yosemite and the California mountains with anecdotes, advice to hikers and geological speculation.

One much neglected but thoroughly delightful and instructive specimen of this practice appeared a century after Carver's departure for London: William H. H. Murray's *Adventures in the Wilderness; or, Camp-Life in the Adirondacks* (Boston, 1869). Its mixture of fact, advice, tall tales and sentiment all loosely assembled is very reminiscent of Carver's *Travels*. There are differences, of course, but these descend not so much from modifications of the model as from the altered role nature reportage had come to occupy in post-Civil War American culture. Murray's book, like Carver's, was popular and is credited with provoking a rush for the wilderness by "Murray's Fools," according to Warder H. Cadbury, author of the Introduction to the recent republication of Murray's *Adventures* (Syracuse, N.Y., 1970). See *American Quarterly*, Annual Review of Books, 1971, p. 323.

18. Could another American, one of giant intellect and talent, or any eighteenth-century author have imagined a fiction that would have synthesized culture and nature as the anthropologists later did? It is implausible, given some understanding of what goes into the origination of expansive new ideas. Intellectual constructs of the magnitude required do not grow *de novo* from any single genius' mind. Or when they appear to have done, they may be said truly to have "come before their time." Certain intellectual developments seem necessary if not sufficient preconditions of such paradigmatic breakthrough. But in addition, the mundane experience of many people must help to make plausible the truth of new propositions about the nature of enculturation and the processes of nature. Too few in Carver's day had experienced the truly staggering variety of exotic cultures; too few had witnessed first hand the incredible malleability of human beings and their ability to "make up" social patterns quite "unnatural" by traditional standards. A century of industrially caused disarray in western culture lent a plausibility to relativistic perceptions of humanity by anthropologists that preindustrial people simply could not share; made plausible for the first time that the species was perhaps more *Homo faber* than *Homo sapiens*. Furthermore, English culture even in all its disarray contained a residue of complex and articulate interconnections such as to cause Englishmen to attempt a new synthesis. American culture seemed to be building, not decaying; improvisation rather than salvage or maintenance, housing rather than architecture were the national preoccupation. Great synthetic geniuses would come from England or the continent, not from America, for years to come. The drama of coping, more than the work of establishment, attracted American genius.

19. James Cook, *Journal of the Resolution's Voyage, In 1772, 1773, 1774, and 1775* (London, 1775), p. iv; Hitchings, Introduction to *John Ledyard's Journal of Captain Cook's Last Voyage* (Corvallis, Oreg., 1963), p. xxi. Three other accounts of interest are John Hawkesworth, *An Account of the Voyages ... Southern Hemisphere* (London, 1773); Sydney Parkinson, *A Journal of a Voyage ... Endeavour ...* (London, 1784); and Joseph Banks, *The Endeavour Journal, 1768–1771* (New South Wales, 1962). Especially fine is the three-volume *A Voyage to the*

Pacific Ocean . . . 1776, 1778, and 1780 (London, 1785), the first two
volumes of which were "written by Captain James Cook, F. R. S."
20. Hitchings, *Ledyard's Journal,* pp. xxxi–xxxv.

Chapter IV

1. Tyler, AHAL, 521–22; Ernest Earnest, *John and William Bartram*
(Philadelphia, 1940), p. 42; Kastner, *A Species,* pp. 65, 76–77, 81.
2. Earnest's biography provides the simplest, and still accurate, biography
of John. Francis Harper's treatment of John Bartram in APST, n.s. 33,
Pt. 1 (1942): 1–122, is an excellent source of biography, bibliography
and notes, and it includes numerous plates in addition to the "Diary
of a Journey Through the Carolinas, Georgia, and Florida from July
1, 1765, to April 10, 1766" (hereafter cited as D). Josephine Herbst's
New Green World (New York, 1954), though irritatingly devoid of footnotes and bibliography, is perhaps the most sensitive to Bartram's delight in his work and travels. Ann and Myron Suttons' *Exploring with
the Bartrams* (New York, 1963), intended apparently for young readers,
includes many pictures and exploits wherever possible the language of
primary sources for the dialogue and descriptions. William Jay You-
mans' account and evaluation of Bartram in *Pioneers of Science in
America* (New York, 1896) records incidents of Bartram's life and
anecdotes about his farming, stonecutting, house building, etc. The
most vivid encounter one can have with Bartram, his life, works,
friends, ideas, sentiments and style can be found in William Darlington's collection of letters to and from Bartram, *Memorials of John
Bartram and Humphry Marshall* (Philadelphia, 1849, rpt. New York,
1967). Crèvecoeur's fictionalized account of a visit to John Bartram by
a Russian traveller, reprinted in Darlington, is considered by one
biographer to be the "story of an actual visit" (Earnest, *Bartram,* p.
14), though all biographers note Crèvecoeur's error in calling John's
father a Frenchman and view with amused tolerance the account of
John's "sudden awakening" to nature's beauties while plowing a field.
3. Bartram's variety of interests is barely suggested by his contributions
to the RSPT. Of the many letters (fifty in Darlington) Bartram sent to
Collinson only eight appeared in the RSPT. (Earnest lists only six, p.
30, missing those after 1751.) See RSPT 41 (1742): 358–59, on "small
Teeth" of rattlesnakes; 43 (1745): 157–59, on "Salt-Marsh Muscles
[mussels]"; 43 (1745): 363–66, on wasps' nests; 46 (1750): 278–79, on
"Black Wasps"; 46 (1751): 323–35, 400–402, for remarkable descriptions
of "Dragon Flies," especially on one coming out of a chrysalis; 52
(1762): 474, on aurora borealis; 53 (1763): 37–38, on the "yellowish"
wasps of Pennsylvania; and 54 (1764): 65–68, on the cicada—a piece by
Collinson with allusions to Bartram's work. Only an index of Darlington could adequately demonstrate Bartram's sweeping interests, and

since a number of items are discussed and identified in the text, no attempt is made to identify by page all the items alluded to in the text. I recommend to those interested in further examples of Bartram's wit, ethos, temper and idiom the following from Darlington: on cranes, M, p. 212; on the rigors of collecting, p. 227; on "aligators," pp. 248–49; on spring, p. 161; on landscape, pp. 194–95; on geology, p. 210; and on climbing pine trees, p. 161. The "Index to Personal Names" and "Plant Name Index" added to the 1967 reprint of Darlington make that edition especially useful.

4. Harper, p. 2; the "Diary" published by the American Philosophical Society in 1942 is a partial account of the 1765–66 journey.

5. For Collinson's tone, see Darlington, M, p. 257; for Bartram's touchiness, p. 215: his ire was particularly aroused by Collinson's 16 June 1757 complaint, "What didst mean, to send me so large a box of seeds? It made much trouble, and time, to part it." "I reflected upon myself," Bartram says, "what pains I had taken to collect those seeds, in several hundred miles' travel, drying, packing, boxing and shipping, and all to put my friend to trouble!" He finished the letter, nevertheless, with good-tempered remarks on his family's affairs. Kastner captures and conveys the human tone of the collective project better than any other (*A Species,* pp. 40–67, et passim).

6. Van Wyck Brooks, *The World of Washington Irving* (New York, 1944), p. 110; Earnest, *Bartram,* p. 15; AHAL, 522.

7. Bartram, *Observations* (London, 1751; rpt. New York, 1974), pp. 11–12. All citations will be to this facsimile of the *Observations.* An earlier reprint (Rochester, 1895) is interesting as a curiosity, bibliographically, in that the compositor chose to render the eighteenth-century long *s* as *f* as well as silently "correct" the text. A modern edition, introduced by Whitfield J. Bell, Jr., and published by the Imprint Society (Barre, Mass., 1973) under the title *A Journey from Pennsylvania to Onondaga in 1743* includes extracts from Lewis Evans's *Journal* and Conrad Weiser's *Report.* The text, Bell acknowledges, "is not a proper facsimile"; typographical errors have been "silently corrected" and spelling and capitalization, regularized; nevertheless, excepting the discordantly modern illustrations, I recommend the book to those interested in Bartram and disinclined to struggle with rare editions, microfilms, or eighteenth-century typography. Bell's brief (twelve pages) but sensitive and informed introduction enhances its worth for instruction and pleasure.

8. My admiration of this passage is not unique. Ann and Myron Sutton, authors of a fictionalized narrative of Bartram's travels, select this incident to open their book, *Exploring with the Bartrams:*

"Look out!"
John Bartram's horse wheeled in the trail.

"Rattlesnake!" he shouted to his companions. "Stay where you are."

A stocky man in a worn leather jacket and rough trousers, Bartram leaped quickly off his horse and grabbed a bleached oak stick.

"You got him, John!"

Interestingly enough, though the Suttons' account is sprinkled with exclamation points and cast as dramatic narrative, Bartram's original is more compelling.

9. *Boswell's Life of Johnson,* Oxford Standard Authors (London, 1960), p. 267, note. RSPT 53 (1763): 37–38; Earnest, *Bartram,* pp. 64–66; Darlington, M, pp. 161–62, 211–13, 234–35.

10. I have tampered with the punctuation of this one passage by eliding thirteen commas. Commas were not a normal ingredient of Bartram's prose. Darlington's renditions of Bartram's letters are surely over-pointed. See Bartram's descriptions in the "Diary" for his customary use of blank spaces to indicate pauses (D, 14, 21, 25). I have made no attempt to "correct" punctuation in other selections from M, but it was important here to remove the commas that obscured the natural rhythms of his speech. In M they occur thus: bright, a; it, when; air, between; study, to; lower, as; lowered, an; garden, where; down, spread; rays, with; breeze, can't; miles, from; meadow, to; plants, they.

11. Earnest, *Bartram,* pp. 48–54; O, ii.

12. Sir Paul Harvey's *Oxford Companion to English Literature* supplies a good, brief biographical paragraph on White; the introduction of *The Natural History of Selborne* (London, 1813), pp. vii–viii, chronicles his life in greater detail. Walter Johnson's "White's Prose Style," in *Gilbert White: Pioneer, Poet, and Stylist* (London, 1928), pp. 271–303, is very complete and perceptive.

13. *Journals of Gilbert White,* ed. Walter Johnson (London, 1931), pp. 276–77.

14. *Indoor Studies,* Vol. 8 of *Writings of John Burroughs* (Boston, 1904), pp. 37–38.

15. Ibid., p. 38.

16. *Walden* in *Works of Thoreau,* ed. Henry Seidel Canby (Boston, 1937), pp. 272, 276, 283–85.

17. Winthrop, "Journal," in Miller and Johnson, eds., *The Puritans,* 1: 137; Thoreau, "Journal," *Works,* p. 604; Howells, "from *Criticism and Fiction,*" in Norman Foerster's *American Poetry and Prose* (Boston, 1957), p. 1088; Steinbeck in *Sea of Cortez,* with Edward F. Ricketts (New York, 1941), quoted by Frederick Bracher, "Steinbeck and the Biological View of Man," *Steinbeck and His Critics* (Albuquerque, 1957), p. 185.

Chapter V

1. *The Natural History of Carolina . . .* (short title, *Carolina*) most often

exists as a two-volume work (one hundred plates in each) with a twenty-plate Appendix at the end of Volume Two. Parenthetical citations refer to Volume One (1), Two (2), or Appendix (App.) of the 1754 London edition; halftone illustrations derive from the 1731–43 London edition, courtesy The Bancroft Library, University of California, Berkeley, California; the color plate, from the 1771 London edition owned by the Minneapolis Athenaeum, Minneapolis, Minnesota. *Carolina* was first published in ten volumes (twenty plates each) and an Appendix (twenty plates); see William T. Stearn, "Publication of Catesby's *Natural History of Carolina*," *J. Soc. Bibl. Nat. Hist.* 3 (1958): 328. Contemporary notice of these volumes can be found in RSPT 36 (1730): 425; 37 (1731): 174; 38 (1734): 315; 39 (1735): 112, 251; 40 (1738): 343; 44: 2 (1747), 599; and 45 (1748): 157. *Carolina* was popular and consequently republished many times both in England and on the continent. Some of the continental pirate editions are quite handsome, though some like *Piscium, Serpentium* ... (Nuremberg, 1777) reproduce only Volume Two. Modern microfilm copies are useful for Catesby's prose and for the general compositional elements of his art, but lack color and blur minute details of engraving. Catesby's *Hortus Britanno-Americanus* ... (London, 1763), according to George F. Frick and Raymond P. Stearns, *Mark Catesby: The Colonial Audubon* (Urbana, Ill., 1961), is less impressive and much of it reduced and copied from *Carolina*. The Beehive Press (Savannah, Georgia) has recently (1972) produced an excellent, though expensive, facsimile of *Carolina,* with selected plates in color.

Gentleman's Magazine 22 (1752): 300, calls *Carolina* "a work which exceeds all others of the same kind for its beauty, accuracy, and splendor; it contains representations of birds, plants, fish, serpents, &c. engraved on more than 200 plates, in a most masterly manner, and coloured from nature with so much elegance as to emulate a painting"; Richard Pulteney, *Historical and Biographical Sketches of the Progress of Botany in England* (London, 1790), 226. For Linnaeus's debt to Catesby see W. L. McAtee, "The North American Birds of Linnaeus," *J. Soc. Bibl. Nat. Hist.* 3 (1957): 291–300. The correspondence of John Bartram (Darlington), Thomas Jefferson, and Benjamin Franklin demonstrate the respect these knowledgeable Americans felt for Catesby's work.

2. Between Frick and Stearns, *Mark Catesby,* and Elsa Guerdrum Allen, "The History of American Ornithology Before Audubon," APST, n.s. 41, Pt. 3 (1951), the student is well supplied with biographical, bibliographical, and critical material relevant to Catesby's scientific contributions. E. P. Richardson, *Painting in America* (New York, 1956), and Oliver Larkin, *Art and Life in America* (New York, 1949), contain brief, general treatments of Catesby and reproduce several plates. Most other one-volume histories of American art ignore Catesby entirely or

mention him only briefly. Some sympathetic criticism and several re-productions of plates are contained in articles published in *Antiques* 37 (1940): 282–84; 52 (1947): 352; 61 (1952): 325–27; and 64 (1953): 322; in *Connoisseur,* 121 (1948): 47–52, and 166 (1967): 124, 126; and in Kastner, *A Species* (see three color inserts and p. 19). Shortly after Catesby's death (1750) *Gentleman's Magazine* began publishing repro-ductions of plates from *Carolina* with the signature "J. C." or "J. Cave":

21 (1751)	10	Rice-birds (Bobolink)		"JC"	*Car.,* 1. 14
22 (1752)	276	Painted finch (Painted bunting)	colored		*Car.,* 1. 44
———		Blue linnet (Indigo bunting)	colored		*Car.,* 1. 45
———	475	Fox-coloured thrush (Br. thrasher)		"JC"	*Car.,* 1. 28
23 (1753)	29	Little thrush		"J. Cave"	*Car.,* 1. 31
———	128	Purple gross beak		"JC"	*Car.,* 1. 40
———	181	Baltimore bird (Oriole)	colored		*Car.,* 1. 48
———	269	Blackcap flycatcher (Phoebe)	colored	"J. Cave"	*Car.,* 1. 53
———	325	Large lark (Meadowlark)	colored	"JC"	*Car.,* 1. 33
———	513	Red bird (Cardinal)	colored		*Car.,* 1. 38
———	608	Rattlesnake	colored		*Car.,* 2. 41.

3. Unless otherwise noted, the biographical material assembled above and in the several paragraphs following comes either from Frick and Stearns, *Mark Catesby,* pp. 3–7, 9, 11–14, 16–17, 19, 23, 25–26, 28, 30, 34–35, 38–43, 50 n. 2, 60, 76–77, or *Carolina,* "Preface."

4. Peter Kalm, *Kalm's Account of His Visit to England on His Way to America in 1748,* trans. Joseph Lucas (New York, 1892), p. 119.

5. Unless necessary to avoid confusion, parenthetical citations to the text and plates of *Carolina* will omit all but the volume and page (or plate) number; e.g., (2: 41) for plate 41 of the second volume, but (App. 11) for the eleventh plate of the Appendix.

6. "Of Birds of Passage," RSPT 44 (1747): 435.

7. Frick and Stearns, *Mark Catesby,* p. 87; *Carolina,* 2: 62; 1: 1, 8, 27.

8. Erwin Panofsky's discussion of the documentary and monumental in art has been more useful in understanding Catesby's art than anything written specifically about Catesby by art historians; see both "The History of Art as a Humanistic Discipline," pp. 10–11, and "Iconog-raphy and Iconology: An Introduction to the Study of Renaissance

Art," pp. 26–54, in *Meaning in the Visual Arts* (Garden City, N.Y., 1955).

9. The relevant portion of Garden's letter (Jan. 2, 1760) to Linnaeus appears both in Frick and Stearns, *Mark Catesby*, p. 76, and in Edmund Berkeley and Dorothy Smith Berkeley, *Dr. Alexander Garden of Charles Town* (Chapel Hill, 1969), p. 133. The full letter appears in Sir James Edward Smith, ed., *Selections of the Correspondence of Linnaeus* (London, 1821), 1: 300–301. Pages 56–57, 75, and 164 in Berkeley and Berkeley further illustrate Garden's dislike of Catesby's work. Jefferson's comments on Catesby occur near the end of Query VI in *Notes on the State of Virginia* (pp. 66–69 in the Harper Torchbook edition, 1964). Interestingly enough, even *Gentleman's Magazine* 21 (1753): 608, which so praised the splendor of *Carolina*'s plates, subtly denied the authority of Catesby's iconography when their engraver freely altered details of composition, implicitly assenting thereby to the subordination of art to reportage; see note 12 below.

10. It would require several pages to catalogue the variety of flora-fauna combinations in *Carolina*, but a rough analysis shows that about 70 percent of the 220 plates feature two or more items (not counting plates in which two species of fish, but no other life, appear); slightly more than 10 percent include extra butterflies, moths, wasps, spiders, or ants in addition to the main plants and animals; about 20 percent (mostly fish, duck, and turtle plates) include nothing beyond the main subject other than seawater, bits of turf, or conventional stumps and perches.

11. Catesby reveals his respect for accurate color, gesture, and ecological juxtaposition when he says in the Preface, "In designing the Plants, I always did them while fresh and just gather'd: And the Animals, particularly the Birds, I painted them while alive, except a very few, and gave them their Gestures peculiar to every kind of Bird, and where it would admit of, I have adapted the Birds to those Plants on which they fed, or have any Relation to" (Preface, xi).

    Albin's book contains nine pictures of American birds; W. L. McAtee lists and identifies these as well as all birds mentioned in Catesby's *Carolina*, whether drawn or simply mentioned, in "The North American Birds of Mark Catesby and Eleazar Albin," *J. Soc. Bibl. Nat. Hist.* 3 (1957): 193–94.

12. Catesby's rattlesnake is handsomely reproduced as one of *Gentleman's Magazine*'s special, hand-tinted engravings, 23 (1753): 608. Unfortunately, the engraver destroyed the symmetry of Catesby's conception by moving all the rattle and fang details over to the right side of the plate. The result is a finely detailed, beautifully colored and thoroughly unbalanced picture—and a persuasive demonstration that Catesby's placement of details in the original functioned to balance the composition as well as inform the viewer.

13. The evolution of the rattlesnake as cultural symbol is beyond the
scope of this chapter, but the topic is suggestive. No other non-useful
American beast intrigued natural philosophers more. Initially the
American rattlesnake had meaning only as a dangerous curiosity, one
of the arcana of nature. Both fabulous and reliable tales became current in the lore of natural philosophy: virtuosi believed with the folk
that rattlesnakes charmed birds and squirrels out of trees. The best
source for anyone wishing to pursue the cultural uses and significance
of rattlesnakes in America is Laurence M. Klauber's two-volume com-
pendium of herpetology and history, *Rattlesnakes, Their Habits, Life
Histories, and Influence on Mankind* (Berkeley, 1956), pp. 217–18, 443–
44, 612, 1188–1260. A massive and admirable accomplishment, Klauber's
book attests the rattlesnake's continuing power to generate and organize the cognitive and affective energies of Americans. The rattlesnake's progress from natural emblem to political device and patriotic
icon can be traced in the literature and art of the eighteenth century.
By mid-century (1751) Benjamin Franklin had advanced the snake's
cultural significance by making it the vehicle of his attack on British
colonization practices in "Felons and Rattlesnakes" (*Papers* 4: 130–33),
a mock proposal that the colonies transport rattlesnakes to London to
counter the Crown's export of felons to the colonies. Franklin's snake
is ambiguous: it is one of Nature's felons, but also a champion of a
righteous colonial cause. He and others linked the snake to the cause
of colonial unity in the following quarter century (1754–76) by pic-
turing a snake chopped into eight or nine parts, each labelled with a
colony's initials, and accompanied by the slogans, *"Join, or Die"* or
"Unite or Die!" *Papers* 5: xiv; Albert Matthews, "The Snake Devices,
1754–1776, and the *Constitutional Courant,* 1765," *Publ. Col. Soc.
Mass.* 11 (1910): 409–53. With the outbreak of hostilities, rattlesnakes
coiled in the center of the "Don't Tread on Me" Gadsden flag and
slithered diagonally upward across the red and white stripes of the
first navy jack; and John Paul Jones designed a uniform for himself
with rattlesnakes on the lapels and buttons (Samuel Eliot Morison,
John Paul Jones, A Sailor's Biography [Boston, 1959], pp. 71–72).

14. Identification by Allen, "American Ornithology," p. 466.

15. See Hulton and Quinn, *American Drawings* . . . (London; Chapel Hill,
N. C., 1964), plate 4 and pp. 50–51. Hulton and Quinn detect six other
cases in which Catesby "borrowed" from White: the iguana, remora,
catfish, gar, puffer (i.e., "globe fish") and swallow tail butterfly, pp. 50–
51. See pp. 7, 27, 67, 70, 73, 82, 127, 130, and 134–35 for further explanation of Catesby's debt to White. See also the paperback edition
of Thomas Hariot's *A Briefe and True Report . . . Virginia,* ed. Paul
Hulton (New York, 1972).

16. Frick and Stearns, *Mark Catesby,* p. 66.

17. *Enci. Univ. Ill. Eur.,* Tomo 54, p. 1258; on Tanje see *Catalogue of Botanical Books in the Collection of Rachel McMasters Hunt,* compiled by Jane Quinby (Pittsburgh, 1961), II, pt. 2, p. 451. Also Thieme et Becker, *Allgemeines Lexikon der bildenden Künstler* (Leipzig, 1911–47). Peter Kalm, *Kalm's Account* (p. 17), calls *Carolina* "precious and costly."

18. Allen, "American Ornithology," pp. 471, 478–80; W. L. McAtee, "The North American Birds of Mark Catesby and Eleazar Albin," *J. Soc. Bibl. Nat. Hist.* 3 (1957): 193–94. See Allen (p. 479) for Albin plate from *Natural History of Spiders.* For a bibliography of bird books see Claus Nissen, *Die Illustrierten Vogelbücher* (Stuttgart, 1953).

19. For an extended and systematic development of the way in which works are immediating, powerful incarnations of being even as they also are culture-specific artifacts, see Armstrong, note 1 (chapter 1) above.

Index

Library of Congress Cataloging in Publication Data
Wilson, David Scofield.
In the presence of nature.
Includes bibliographical references and index.
1. Natural history—Addresses, essays, lectures.
I. Title.
QH81.W489   500.9   77-90733
ISBN 0-87023-020-4